Synchrotron Radiation - Useful and Interesting Applications

Edited by Daisy Joseph

Published in London, United Kingdom

IntechOpen

Supporting open minds since 2005

Synchrotron Radiation - Useful and Interesting Applications
http://dx.doi.org/10.5772/intechopen.79284
Edited by Daisy Joseph

Contributors
Amardeep Bharti, Navdeep Goyal, Philippe Sciau, Tian Wang, Elena Ivanova, Samuel Cheeseman, Khanh Truong, Jitraporn Vongsvivut, Mark Tobin, Russell Crawford, Daisy Joseph

First published in London, United Kingdom, 2019 by IntechOpen
IntechOpen is the global imprint of INTECHOPEN LIMITED, registered in England and Wales, registration number: 11086078, The Shard, 25th floor, 32 London Bridge Street London, SE19SG – United Kingdom
Printed in Croatia

British Library Cataloguing-in-Publication Data
A catalogue record for this book is available from the British Library

Additional hard copies can be obtained from orders@intechopen.com

Synchrotron Radiation - Useful and Interesting Applications
Edited by Daisy Joseph
p. cm.
Print ISBN 978-1-83880-441-1
Online ISBN 978-1-83880-442-8
eBook (PDF) ISBN 978-1-83880-646-0

Meet the editor

Dr. Daisy Joseph has a PhD in Physics on the topic "Study of X-ray spectra by photo, proton and heavy ion excitation." She is an expert in EDXRF and PIXE. She has also carried out XRF by synchrotron: by XANES and EXAFS, measuring X-ray absorption edges using the Indus-2 synchrotron, RRCAT, Indore, India. She has around 100 publications in peer-reviewed journals and is the associate editor for the *Journal of Advances in Basic Sciences* and the *Bioscience* journal. She has guided several PhD students from India. She has represented BARC, India, at a large number of conferences and workshops in her field. Dr. Joseph is a collaborator for the following institutes in India: Christ University, Bangalore; Karnatak University, Dharwad; Punjab University, Chandigarh; SR and BGR College of Science and Arts, Khammam, Andhra Pradesh; National Institute of Manipur, Langol, Manipur; University Department of Physics, Mumbai University; and BITS Pilani, Rajasthan. She has reviewed articles in the following journals: *X-ray Spectrometry*, *Journal of Radiation Isotopes*, and the *Bioscience* journal. Dr. Joseph has also been a reviewer for DAE Solid State Physics Symposium and for Spectroscopy Letters.

Contents

Preface

A huge amount of research has been directed at an upcoming modern technique in X-ray physics—synchrotron radiation. A large number of books, journals, and articles covering its basics, methods of experimentation, and applications have thrown light on its usefulness, versatility, and excellent performance. Hence, it seems logical to give all scientists, including physicists, chemists, biologists, archeologists, and medical professionals, the opportunity to read further interesting applications in this exciting field of research. The book consists of four chapters covering applications in physics, cultural heritage, and bioanalytics.

The first chapter covers the basic concepts of synchrotron light and its implications. The second chapter covers the fundamentals of synchrotron radiation, basic experimentation in physics, and scope of applications. The third chapter covers the useful application of synchrotron light to unravel the mysteries of cultural heritage with reference to ceramics, and the fourth chapter has an interesting application for the study of insect wings.

The book will make for thought-provoking reading because the chapters are crisp and novel in their approach. I hope this book will enrich the readers in all basic sciences, including students and professionals, to become motivated in carrying out better research in all or similar fields of synchrotron radiation and allied areas.

Dr. Daisy Joseph
Nuclear Physics Division,
Bhabha Atomic Research Centre,
Trombay, Mumbai, India

Chapter 1

Introductory Chapter: Synchrotron Radiation-Basics and Concepts

Daisy Joseph

1. Introduction

When an electron traveling at nearly the speed of light in an orbit, emits a continuum of electromagnetic radiation tangential to the orbit, it gives you a synchrotron light which is the synchrotron radiation.

The main difference is that a cyclotron accelerates the particles in a spiral since the magnetic field is constant, whereas the synchrotron adjusts the magnetic field to keep the particles in a circular orbit. There are now more than 60 synchrotrons and free electron lasers (FELs) around the world dedicated to applications in physics, engineering, pharmacology, and new materials, to name but a few. As the electrons are deflected through the magnetic field created by the magnets, they give off electromagnetic radiation, so that at each bending magnet, a beam of synchrotron light is produced. SR—synchrotron radiation—can be used in a variety of spectroscopy techniques, namely, XAFS, soft X-ray, imaging, X-ray lithography, dispersive EXAFS, scanning EXAFS, EDXRD, XRF, protein crystallography, and X-ray beam diagnostic visible beam diagnostic to name a few.

1.1 Synchrotron radiation can give us a versatile field of X-ray spectroscopy which is the X-ray absorption spectroscopy (XAS)

X-ray interacts with all electrons in matter when its energy exceeds the binding energy of the electron. X-ray excites or ionizes the electron to a previously unoccupied electronic state (bound, quasi bound, or continuum). The study of this process is **XAS**.

Since the binding energy of core electrons is **element specific**, XAS is element and core level specific (e.g., Si K-edge at 1840 eV is the 1s electronic excitation threshold of silicon).

X-ray Absorption spectroscopy is often referred to as follows:

NEXAFS for low Z elements (C, N, O, F, etc. K-edge, Si, P, S, L-edges) or

XAFS (**XANES** and **EXAFS**) for intermediate Z and high Z elements.

As core electron is excited with *hν* greater than or equal to the threshold (E_o), it is excited to a final state defined by the **chemical environment**, which modulates the absorption coefficient relative to that of a free atom. This **modulation** is known the **XAFS** (**Figure 1**),

- XAFS contains all the information about the local structure and bonding of the absorbing atom

- XAFS study requires a **tunable X-ray source** which is the synchrotron radiation

IntechOpen

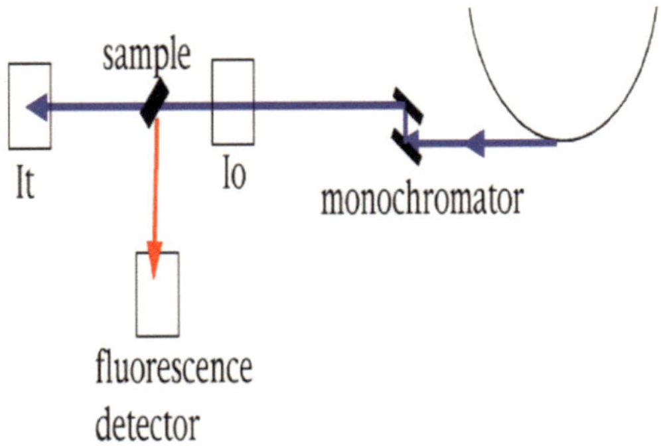

Figure 1.
XAS instrumentation.

NEXAFS (near edge X-ray absorption fine structures) describes the absorption features in the vicinity of an absorption edge up to ~50 eV above the edge (for low Z elements for historical reasons). It is exactly the same as **XANES** (X-ray absorption near edge structures), which is often used together with **EXAFS** (extended X-ray absorption fine structures) to describe the modulation of the absorption coefficient of an element in a chemical environment from below the edge to ~50 eV above (XANES), then to as much as 1000 eV above the threshold (EXAFS). **NEXAFS** and **XANES** are often used interchangeably. **XAFS** and **XAS** are also often used interchangeably.

Molecular spectra in absorption can be measured very efficiently using SR, and a survey of molecular cross-sections in the XUV of importance for Earth's atmospheric processes is at present being conducted. The energy of the electrons ejected by the photons can be measured with high accuracy. The binding energy of electrons in atoms is thus measured directly. The technique is generally called photoelectron spectroscopy (PES). The code symbols XPS, UPS, and ESCA are used to differentiate applications:

XPS is PES with X-ray sources; UPS is PES with U.V. sources. ESCA means electron spectroscopy for chemical analysis.

I hope that this book on Synchrotron Radiation will elucidate the technique of synchrotron radiation in a very crisp and precise manner. It will definitely bring out not only the technological aspects but also its applications in a variety of fields.

Author details

Daisy Joseph
Nuclear Physics Division, Bhabha Atomic Research Centre, Trombay, Mumbai, India

*Address all correspondence to: djoseph@barc.gov.in

Chapter 2

Fundamental of Synchrotron Radiations

Amardeep Bharti and Navdeep Goyal

Abstract

Synchrotron radiations are emerging as a real-time probing tool for the wide range of applied sciences. Synchrotron radiations have unique properties because of their high brilliance, collimations, broad energy spectrum, and coherence power that break the limits to characterize the material properties than previous laboratory-based tabletop sources. The third-generation synchrotron light sources are capable of producing 10^{12} times higher brilliance than laboratory-based sources using insertion devices. In this chapter, the fundamental aspects of synchrotron radiations and their generation process have been discussed. The effect of insertion devices and the double-crystal monochromator (DCM) toward the X-ray beam optics has been also discussed.

Keywords: synchrotron radiation sources, insertion devices, double-crystal monochromator, brilliance, X-ray optics

1. Introduction

The synchrotron is basically a cyclotron in which relativistic charged particles are forced to follow curved trajectories under applied magnetic fields, and due to such motion, they emit electromagnetic radiations (infrared to hard X-rays) known as synchrotron radiations [1, 2]. Synchrotron radiations were first observed in 1947 at General Electric particle accelerator, USA, but were considered a nuisance because they caused the loss in particle energy and treated as a particle physics problem [1]. In the 1960s, they were accepted as an exceptional property of light that overcame the shortcomings of X-ray tubes [1, 3]. The exponential growth in the use of synchrotron radiation began after realizing its importance in condensed matter physics. At present, synchrotron radiations are widely used for the structural analysis of the matter, from the surface of solids to protein molecules [4, 5].

A synchrotron is composed of five main components: electron source, booster ring, storage ring, RF (radio-frequency) supply, and beamlines. In general, electrons are generated by the thermionic emission from a hot filament (electron gun) which serves as a source [1–3]. The electrons are then accelerated by either microtron or linear accelerator (LINAC) to several hundred MeV of energy.

The electrons are then injected to a circular accelerator to boost its energy to approach main storage ring electron energy, called booster ring. Electrons are periodically transferred to storage rings from booster when storage ring current falls to $1–1/e \sim 70\%$ to maintain beam current [3].

The storage ring is the main component of a synchrotron, in which electrons travel in a closed path under the effect of magnetic field. The main magnetic components are a set of magnets: bending (dipole), quadruple and sextuple magnets, so-called magnet lattice. Bending magnets force the electrons to follow the closed path, and the beam is focused by quadruple magnets via compensating electronic coulomb repulsion. Sextuple magnets serve as a corrector for chromatic aberration governed from quadruple focusing [1–4].

In the storage ring, the electrons travel at relativistic (99.999%c) speed and kinetic energies of the order of GeV. The modern ring structure consists of a periodic arched section having bending magnets and straight sections composed of insertion devices that are used to produce intense synchrotron radiations, which are named as the "third-generation storage ring" [4] as shown in a **Figure 1**.

As the electrons are accelerated in a circular path, they radiate at frequencies in the visible, ultraviolet, and X-ray regions of the spectrum and lose their energy. A radio-frequency (RF) cavity supplies suitable amount of energy each time the electrons pass through it, to prevent the electron scattering from an inner wall of the storage ring [1–4].

The beamlines are working along the axes of insertion devised and tangential to bending magnet and the storage ring. The beamlines are designed for specific dedicated applications, i.e., X-ray imaging (tomography), X-ray absorption spectroscopy (X-ray absorption fine structure (XAFS), near-edge and extended-edge spectroscopy (XANES, EXAFS)), X-ray scattering (small- and wide-angle X-ray scattering) and X-ray fluorescence/emission spectroscopy (XRF).

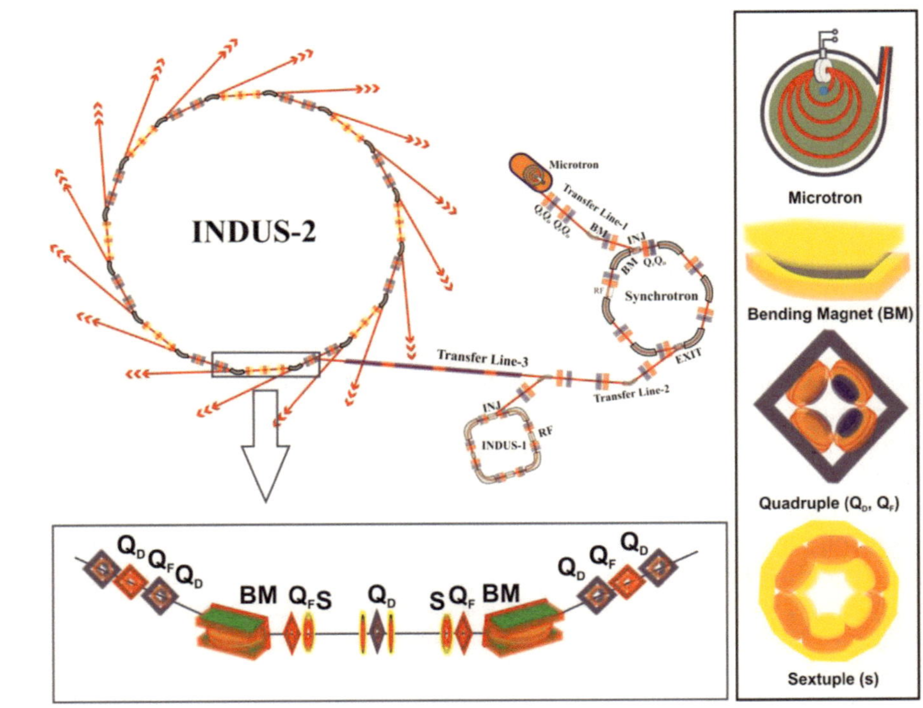

Figure 1.
Detailed outer structure of RRCAT—Indian third-generation synchrotron light source. Electrons are generated via microtron and transferred to the synchrotron for making an accelerated e^--beam. The energetic e^--beam has been introduced to the Indus-1 and Indus-2 via transfer lines. The beam optics have been maintained by the periodic repetition of several magnetic components (quadruple focused (Q_F), quadruple defocused (Q_D), sextuple (S), and bending magnets (BM)) to force e^- beam to follow a circular path.

2. Synchrotron radiation properties

Synchrotron radiations are highly polarized, are intense, and have the broad spectral range from infrared to hard X-ray region. These properties are tunable and can be understood by classical electrodynamic laws [1]. When a charged particle (electron) is moving in a circular path having energy much less than its ground state energy (mc^2 = 0.51 MeV), it behaves as weak dipole which radiates isotropically. In case of relativistic energy, the radiated energy folded in the forward direction makes a narrow cone with a solid angle $\psi \sim \gamma^{-1} = mc^2/E$ as shown in a **Figure 2**. Due to the collimation of radiated photons in such a small angle, cone results in highly intense beam and power radiations in forwarding direction even at distance of tens of meters from storage ring [1–3].

According to special theory of relativity, the kinetic energy of a particle having rest mass m moving with velocity v is given ≪ by [5]:

$$
\begin{aligned}
E &= \frac{\mathrm{mc}^2}{\sqrt{1-\left(\frac{\mathrm{v}}{\mathrm{c}}\right)^2}} \\
&= \frac{mc^2}{\sqrt{1-(\beta)^2}}
\end{aligned} \tag{1}
$$

$$
\gamma = \frac{1}{\sqrt{1-(\beta)^2}} \tag{2}
$$

where $\beta = v/c$ and γ is relativistic correction factor. If the particles approach the speed of light, their mass becomes multiple of factor γ. The particle would have gained infinite mass at $v = c$, which means that the β is a prime factor to consider for the synchrotron radiation. From the above relation, we can write β as

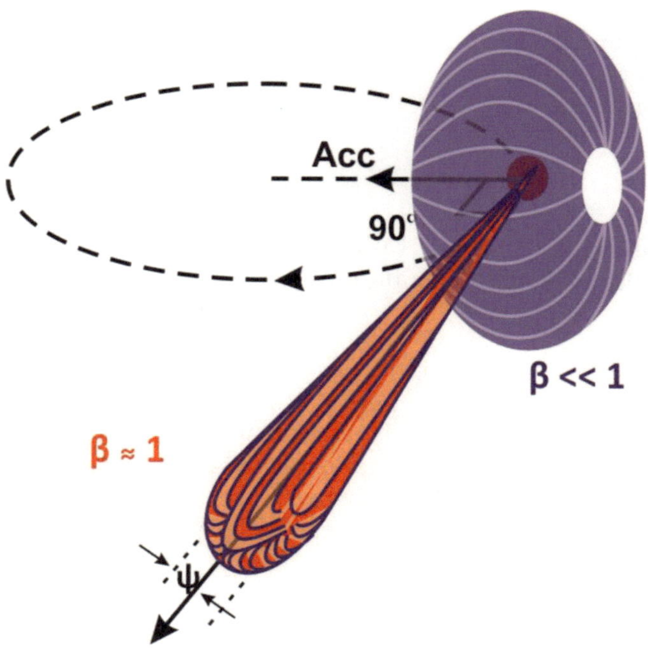

Figure 2.
Radiation pattern of charged particles moving in a circular path: blue ($\beta \ll 1$) and red ($\beta \approx 1$).

$$\beta = \left(1 - 1/\gamma^2\right)^{1/2} \quad (3)$$

For the case of relativistic energy, $1/\gamma^2$ is too small, and the above Taylor expansion can be solved by ignoring the higher terms as

$$\beta \approx \left(1 - \frac{1}{2\gamma^2}\right) \quad (4)$$

In the storage ring, electrons are traveling in a closed path under the effect of the magnetic field which applies a Lorentz force normal to the motion of the charged particle and the magnetic field vector. As the orbital radii (r) remain constant, we can equate the Lorentz force to centripetal force under classical regime as

$$e\,\vec{v} \times \vec{B} = \frac{mv^2}{r} \quad (5)$$

In case of $\beta \approx 1$, m and v must be replaced by the relativistic mass γm and c, respectively. Therefore,

$$ecB = \frac{\gamma mc^2}{r} \quad (6)$$

$$\Rightarrow \mathrm{r} = \frac{\gamma mc}{\mathrm{eB}} = \frac{E}{ecB} \quad (7)$$

So, in terms of standard synchrotron practical units, the above equation can be rewritten as

$$r[m] = 3.3\,\frac{E[GeV]}{B[T]} \quad (8)$$

This implies that the magnetic lattice and storage ring's radius set the limits to the beam energy. The characteristic frequency Ω_c of a synchrotron source in terms of γ and Ω_0 is given as

$$\Omega_c = \frac{3}{2}\gamma^3\Omega_0 = \frac{3}{2}\gamma^3\frac{c}{r} \quad (9)$$

Using the above formulations, we can write the critical energy of a synchrotron light source which is exactly half of the total power emission from a bending magnet as

$$\hbar\Omega_c[keV] = 0.665\,E^2[GeV]B[T]. \quad (10)$$

There are basically two important quantities which can be used to characterize the properties of emitted beam from various light sources: flux and brilliance. The flux is defined as the total number of photons per second per unit *0.1*\% bandwidth radiated in unit angular spread θ along the orbital plane and integrated over the vertical ψ opening angle [1–5]:

$$Flux = \frac{d\phi}{d\phi}\ [\mathrm{photons/s/0.1\%bw/mrad}\ \theta]$$

$$Brilliance = d^4\phi/d\theta d\psi dxdz\ \left[\mathrm{photons/s/0.1\%bw/mrad^2}\right] \quad (11)$$

Brilliance/brightness of the emitted radiation beam has been enhanced over the years as the technology evolves as shown in **Figure 3**. The synchrotron sources enhanced the brilliance of radiations 20 times than that of X-ray tubes/ laboratory-based sources. Due to such a high brilliance, synchrotron radiations are able to explore the deep-inside of materials properties and of utmost interest of scientific research community. In third-generation synchrotron source, both of these quantities can be tuned according to the experimental requirement. The important parameters of the world's top third-generation synchrotron light sources in decreasing order of their storage ring energy are listed in **Table** 1 [2, 4, 6]. Insertion devices have been installed in the storage ring to enhance the flux and brilliance of photon beam.

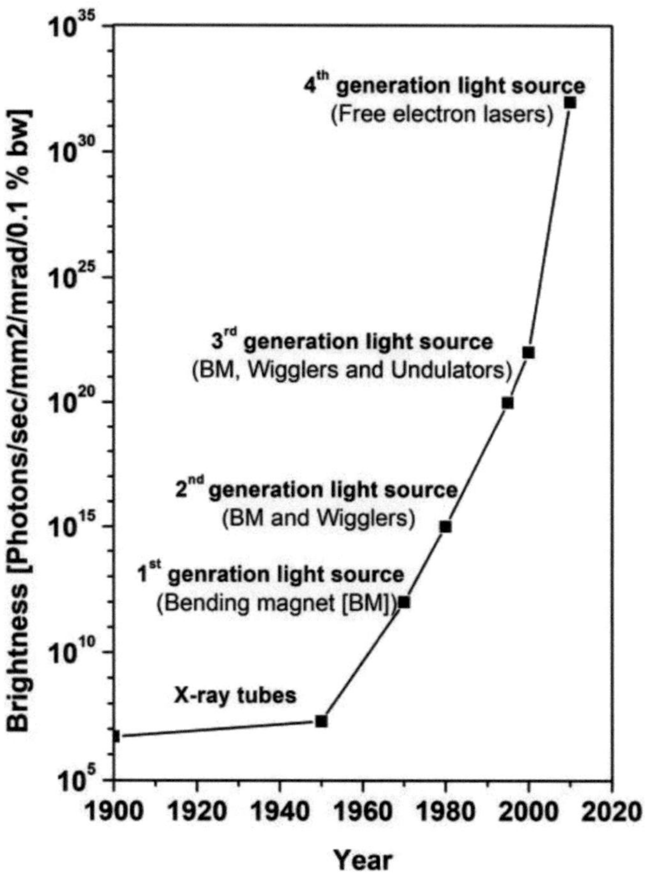

Figure 3.
Evolution of brightness of emitted radiations from the different technical sources over time.

Facility	Country	Energy [GeV]	Current [mA]	Periphery [m]	Emittance [nm-mrad]	Brilliance [ph/s/ mrad^2/mm^2/0.1%bw]
Spring8	Japan	8.00	100	1436.00	2.80	2.0×10^{21}
APS	USA	7.00	100	1104.00	3.00	8.0×10^{19}
PF-AR	Japan	6.50	60	377.00	294.00	4.0×10^{13}
ESRF	France	6.00	200	844.00	3.80	8.0×10^{20}
PETRA-III	Germany	6.00	100	2304.00	1.00	2.0×10^{21}

Facility	Country	Energy [GeV]	Current [mA]	Periphery [m]	Emittance [nm-mrad]	Brilliance [ph/s/ mrad^2/mm^2/0.1%bw]
CHESSS	USA	5.29	200	768.43	101.70	10^{15}
SSRF	China	3.50	300	432.00	3.90	10^{20}
Candle	Armenia	3.00	350	216.00	0.08	1.7×10^{13}
AS	Australia	3.00	200	216.00	7.00	4.6×10^{18}
Diamond	England	3.00	300	562.00	2.70	3.0×10^{20}
SLAC	USA	3.00	500	234.00	0.01	2.0×10^{13}
PAL	South Korea	3.00	400	281.82	5.80	2.0×10^{11}
ALBA	Spain	3.00	100	268.80	4.58	4.0×10^{12}
NSRRC	Taiwan	3.00	500	518.40	1.60	10^{16}
NSLS-II	USA	3.00	500	792.00	0.60	3.0×10^{21}
CLS	Canada	2.90	250	171.00	18.10	1.5×10^{11}
SOLEIL	France	2.75	500	354.00	3.70	10^{20}
ANKA	Germany	2.50	200	110.00	50.00	10^{18}
INDUS-II	India	2.50	200	110.00	50.00	10^{13}
PF	Japan	2.50	450	187.00	36.00	3.0×10^{14}
Elletra	Italy	2.40	320	260.00	7.00	10^{19}
SLS	Switzerland	2.40	400	288.00	5.50	5.0×10^{15}
BSRF	China	2.20	100	240.40	7.60	7.0×10^{12}
MAX-IV	Sweden	2.00	500	528.00	0.17	2.2×10^{21}
ALS	USA	1.90	100	240.00	6.80	3.0×10^{18}
BESSY-II	Germany	1.70	100	240.00	6.00	5.0×10^{18}
SAGA	Japan	1.40	300	75.60	25.00	10^{13}

Table 1.
Important parameters of the world's top third-generation synchrotron light sources, tabular in order to decrease storage ring energy [2, 4, 6].

3. Bending magnet/superbend source

The bending magnets (BM) in the storage rings are the primary sources of radiations. Although the primary aim of the (BM) is to circulate the e^--beam in a closed path, as an e^- is forced to move in an arc through the magnetic field by BM, it produces radiation in a flattened cone [1–3]. The BM source produces a beam of fixed vertical opening angle $\psi \sim \gamma^{-1}$ (photon beam divergence), while the horizontal spam is determined by the length of the BM arc. The critical energy of a synchrotron source depends upon the storage ring energy and the magnetic field of BM. The flux spectrum of a source of BM having different magnetic field strength is shown in **Figure 4**. But the technology imparts limits to field strength of permanent magnets ≈ 1 T; only electromagnets made up of superconductor materials (niobium alloys) under cryogenic temperature (liquid He cooling) can provide the field

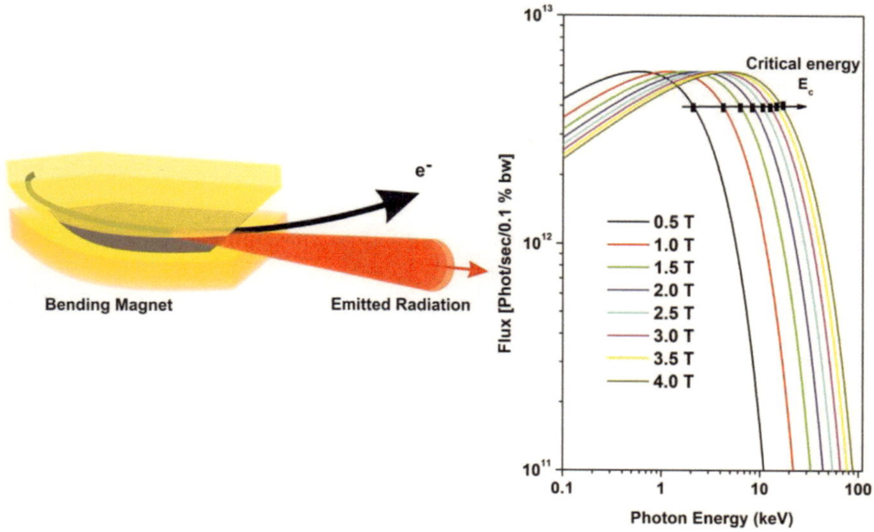

Figure 4.
Pictorial representation of generation of X-ray beam from bending magnet source and the comparative change in the flux spectrum as a function of BM field strength. Graph shows the enhancement in the critical energy with magnetic field strength.

strength of ~ 5 T [2, 7], called superbend source. In this manner, the critical energy of a synchrotron source can be enhanced to 4–5 times.

4. Insertion devices

Insertion devices are the magnetic devices which can supply periodic magnetic fields on the e^--beam through the straight section of the storage ring [1]. The so-called terminology “insertion devices” is used due to its characteristic role, which can be added or replaced from drift space between two bending magnets without perturbing normal operation. There are basically two types of insertion devices: (1) wiggler and (2) undulator, which have been designed to enhance the characteristics (flux, brilliance, energy) of the radiated photon beam [1–3, 6].

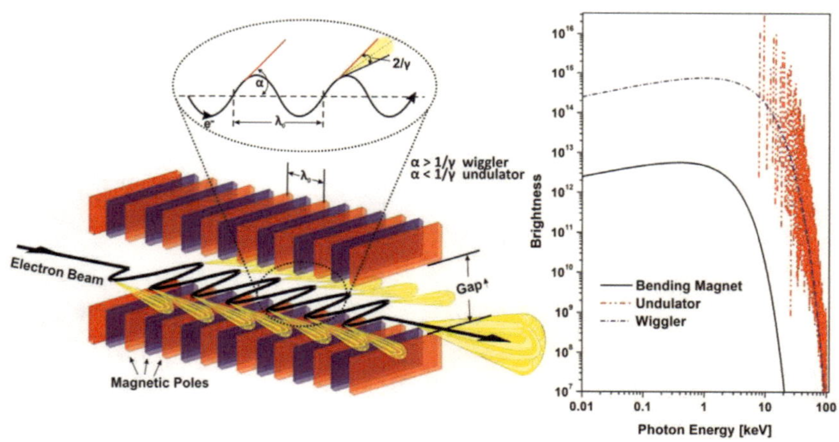

Figure 5.
Layout of radiation beam emission from permanent multipole wiggler/undulator insertion devices used in most of the third-generation synchrotron light sources. The brightness spectrum of the emitted radiations from the individual device source has been shown.

A wiggler is a multipole magnet having periodic (N) arrangement of length (λ), through which it forced an electron to periodically wiggle around its natural path as shown in **Figure 5**. Each wiggle acts as a bending magnet source, so the superposition of radiation from each wiggle enhances the flux and brilliance which is directly proportional to the number of poles N and corresponding magnetic field strength [6]. The horizontal spam of wiggler radiation is defined by $K\gamma^{-1}$, where K is the magnetic deflection parameter and defined as the ratio of wiggling angle (α) of trajectory to the angular aperture (1/γ) given as

$$K = \alpha\gamma \tag{12}$$

An undulator is working on the same principle of wiggler, but $\alpha \leq \gamma^{-1}$ and $K \ll 1$ as shown in **Figure 5**. In this condition, radiated photon loops overlap and interfere with each other along the trajectory. Due to the interference effect in the emitted radiations by individual poles in undulator, it produces the pseudo-monochromatic energy bands called harmonics.

5. Double-crystal monochromator

Synchrotron radiations deliver many features (continuous energy spectrum, high flux, highly collimated and polarized radiation) which are of intense concern in X-ray experimentations. In order to employ this radiation source for the broad area of X-ray scattering/absorption experiments, one has to optimize the various parameters of synchrotron radiations [5]. As for diffused scattering, high flux and low-energy resolution beam are required, while for inelastic scattering high-energy resolution is required. Synchrotron radiations are highly tunable and can be optimized by proper selection of optical design in the beamlines [3, 6].

In condensed matter physics/material science, most experiments required a wide energy range, high-energy resolution/precision, high flux, and focusing. In particular, a monochromator plays an important role to tune the energy of radiation emitted by BM, wiggler, and undulator [7, 8]. BM and wiggler radiate a continuous energy spectrum known as a white beam, while undulator produces pseudo-monochromatic energy bands.

In general, most of the monochromators have two-bounce geometry called double-crystal monochromator [3]. In this geometry, the first crystal diffracts the energy spectrum as a function of incident angle and monochromatizes the synchrotron radiation, while the second crystal adjusts the beam height and direction as

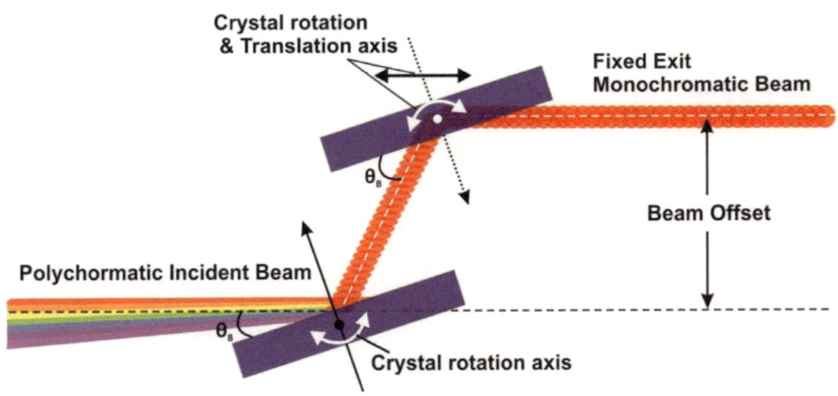

Figure 6.
Geometry of fixed-exit double-crystal monochromator (DCM) system.

shown in **Figure 6**. The energy tuning can be achieved by the crystal's plane rotation and horizontal focusing by sagittal bending of the second crystal [2].

The choice of the crystal is favored by the available crystal quality, thermal conductivity, and its ability to resist from radiation damage [3]. Silicon crystal is widely used due to its relatively high abundance, high thermal conductivity, and mass production as a perfect single crystal of tens of centimeter because of its intense use in the semiconductor industry. As the energetic radiations fall on the crystal, lose energy, and make surface hot, the liquid nitrogen cooling is required to minimize the mechanical strain prompt by incident radiations.

6. Conclusions

In this chapter, the generation of emitted radiations from the different sources has been discussed. The effect of individual source on the emitted radiation brilliance has been studied and provides the understanding of choosing the correct source for particular applications. Such high brilliance of synchrotron light sources makes it a likely/suitable candidate for real-time investigation of materials as compared to that of laboratory-based/X-ray tube sources.

The synchrotron beamlines are dedicatedly designed for specific applications, i.e., X-ray imaging (tomography), X-ray absorption spectroscopy (X-ray absorption fine structure (XAFS), near-edge and extended-edge spectroscopy (XANES, EXAFS)), X-ray scattering (small- and wide-angle X-ray scattering), and X-ray fluorescence/emission spectroscopy (XRF). Moreover, a single beamline can be utilized as a simultaneous multi-characterization tool that would definitely explore the insight of new physics of devices.

This works gives an insight to the basic understanding of the generation of synchrotron radiations and its tunable characteristics via insertion devices which is the fundamental requirement of X-ray spectroscopy.

Acknowledgements

Authors thank the IntechOpen publishers for inviting us to write the book chapter. We acknowledge the University Grants Commission (UGC) for providing the fellowship under basic science research (UGC-BSR) scheme.

Author details

Amardeep Bharti* and Navdeep Goyal
Department of Physics, Panjab University, Chandigarh, India

*Address all correspondence to: abharti_phy@yahoo.com

References

[1] Balerna A, Mobilio S. Synchrotron Radiations: Basics, Methods and Applications. Germany: Springer; 2015; **1**:3-27

[2] Willmott P. An Introduction to Synchrotron Radiation: Techniques and Applications. United States: John Wiley & Sons; 2011. pp. 39-71

[3] Shenoy G. Basic characteristics of synchrotron radiation. Structural Chemistry. 2003;**14**:3-14. DOI: 10.1023/A:102165672

[4] Schlachter AS, Wuilleumier FJ. New Directions in Research with Third-Generation Soft X-ray Synchrotron. Germany: Springer; 1994

[5] Bilderback DH, Elleaume P, Weckert E. Review of third and next generation synchrotron light sources. Journal of Physics B: Atomic, Molecular and Optical Physics. 2005;**38**:S773-S797. DOI: 10.1088/0953-4075/38/9/022

[6] Duke PJ. Synchrotron Radiation: Production and Properties. United Kingdom: Oxford University Press; 2000

[7] Robin D et al. Superbend project at the advanced light source. In: Proceedings of the 2001 Particle Accelerator Conference (IEEE); 18–22 June 2001; Chicago, IL, USA: IEEE; 2002. pp. 2632-2634

[8] Cerino J, Stohr J, Hower N, Bachrach RZ. An ultra-high-vacuum double crystal monochromator beam line for studies in the spectral range 500–4000 eV. Nuclear Instruments and Methods. 1980;**172**:227-236. DOI: 10.1016/0029-554X(80)90639-4

Chapter 3

Full-Field Transmission X-ray Microspectroscopy (FF-XANES) Applied to Cultural Heritage Materials: The Case of Ancient Ceramics

Philippe Sciau and Tian Wang

Abstract

Synchrotrons provide more and more significant analytical techniques to investigate ancient materials from cultural heritages. New ways to visualize the complex structure of these materials are developed on the basis of elemental, density, and refraction contrasts. The tunability of synchrotron beams owing to the high flux and high spectral resolution of photon sources is at the origin of the main chemical speciation capabilities of synchrotron-based techniques. Among them the full-field X-ray absorption near-edge structure (XANES) imaging technique using hard X-rays is particularly efficient. It allows investigating a significant volume of material with a very good spatial resolution, which is invaluable for ancient material because of their heterogeneity and complexity. After presenting the technique and its variants, we will show its ability to study cultural heritage materials through a few examples.

Keywords: FF-XANES, cultural heritage material, attic ceramic, Roman pottery, Chinese porcelain, Qinghua

1. Introduction

Recent decades have seen the tremendous increase in the use of synchrotron radiation-based techniques to study cultural heritage materials [1, 2]. The materials involved in ancient artifacts are very diverse including organic and inorganic compounds such as skins, textiles, woods, resins, stones, metals, alloys, glass, ceramics, or concretes to name only the most current. However, despite this great diversity, most materials used in the design of ancient objects have in common a significant complexity and heterogeneity compared to standard modern manufactured materials. This complexity is not only due to the chemical composition, which can involve a lot of elements, but especially comes from the spatial distribution and organization of these elements leading often to very complex structures. In fact, many of the materials developed by ancient civilizations can be assimilated to composite materials made of a large number of components presenting significant variations in composition, size, or shape. The spatial distribution of components is not so regular as in standard modern materials but it is not random. Often, the components distribution shows a rather nice organization, which can even lead to partially hierarchized structure, which often plays a key role in the physical properties. In addition, the structuration of the components distribution at different scales from nanometer to millimeter is an invaluable source of information about the manufacturing process. This is a very important point since it is often the only main information source of manufacturing technology used for significant productions such as Chinese porcelain, Attic ceramic, *Terra Sigillata*, or Roman concrete to name a few. The equipment used to manufacture them was destroyed, and the structural remains discovered during excavations are often so degraded that it is very difficult to deduce some information concerning the facility operation. A few historical texts mentioning technical processes have been preserved until today. Unfortunately, these valuable documents are not always detailed or completed and moreover, often not written by manufacturers or users. To be really efficient, the study of these documents must be associated to physicochemical analyses of materials produced by the mentioned technologies. Actually, since the main remains left by ancient technologies are the produced materials, a reverse engineering-type investigation is the often best way to obtain pertinent data about manufacturing process [3].

A scientific discipline, called "archaeometry," is dedicated to the analysis of ancient artifacts by physicochemical ways. Many different techniques are used in routine, and for a few decades, a strong growth of nonevasive and portable techniques has been observed for analyzing materials of museum objects on site, without any damage. However, the analytical techniques requiring sampling or sample preparation continue to be developed, in particular the techniques allowing us to characterize the chemical composition and the structure at different scales such as the scanning and transmission electron microscopies (SEM and TEM) and of course the synchrotron radiation-based microscopies that we will present in detail in the next paragraphs. Materials constituting valuable artifacts can be found in many less precious pieces from which suitable tiny fragments can be sampled.

Synchrotron beamlines are, of course, nonportable facilities, but they can be used to directly analyze objects without preliminary sampling and without damage insofar as the analyzed area is not altered by the beam. However, if one wants to take full advantage of the latest developments in terms of high spatial resolution such as submicrometer to nanometer resolutions, it is often necessary to prepare samples with suitable sharps and sizes. These advanced technical platforms, specially beamlines combining X-ray imaging and spectromicroscopy such as ID21 at the European synchrotron radiation facility (ESRF) in Grenoble (France) for instance, are perfectly suitable for the detailed study of complex structure of ancient materials [2]. The large field of view (FOV) combined with a high resolution allows for characterizing the structural organization on large scale, typically from a few millimeters to a few hundred nanometers. Spatial resolution of a few dozen nanometers can be reached on some beamlines with a small decrease in the field of view. Up to now, nanometer resolutions are not yet available and require transmission electron microscopy. However, in many cases, the investigated scale range is large enough to provide a fairly detailed structural description allowing to deduce significant data concerning the manufacturing processes or the relationships between structure and physical properties [4]. The large-scale range also allows us to study some scale laws governing the structure, to identify the more pertinent scales according to a defined physical property and other parameters involved in the study of complex hierarchized materials. If the elemental composition can be studied at high resolution on the majority of micro- and nanobeamlines taking into account the beamline energy range of line and the targeted elements, it is different for the mineral or crystalline phase composition. Structural composition can be determined from X-ray diffraction (XRD) or deduced from X-ray absorption spectroscopy (XAS). However, even if more and more beamlines propose the two techniques, they are seldom optimized for both. A choice must be done between the two approaches as much as the diffraction requires associating monochromatic (powder) and polychromatic (single crystal) diffractions to be efficient. In XRD experiments, the spatial resolution is given by the size of the X-ray beam. With a beam size of a few microns or smaller, only the crystallites with sizes very inferior to the beam size give a usable diffraction pattern in monochromatic mode; X-ray powder diffraction conditions. The study of larger crystals relative to the beam size requires polychromatic beam in order to record a large number of reflection on the 2D diffraction pattern to be able to identify the nature and the orientation of the crystallite (or a few crystallites) irradiated by the beam; X-ray Laue diffraction conditions. This coupled diffraction approach applied to the cultural heritage materials is detailed in [5]. In this chapter, we will present only the approaches based on X-ray absorption spectroscopy, which currently give the best spatial resolutions and are much less time-consuming.

2. X-ray full-field XANES analysis

Usually, the high spatial resolution is obtained by focusing the X-ray beam on the sample surface as much as possible in order to have the smallest probed area, for instance, the smallest irradiated area, on ID21 (ESRF), down to a rectangle of around 300 nm (V) × 700 nm (H), using a focusing Kirkpatrick-Baez mirror system. The irradiated area can be decreased for monochromatic X-rays using zone plates ("ZP") at the expense of flux, chromaticity, and beam stability [2]. Detectors positioned around the sample allow for recording transmitted beam intensity and the X-ray fluorescence (XRF) emission. µXRF 2D-mappings are obtained by raster-scanning the sample over areas of interest and recording the signals with different detectors. Element maps are quickly obtained from the XRF emission. Punctual X-ray absorption spectra (XAS) can be easily obtained directly from the measurement of the intensity of the transmitted beam versus the energy of the X-ray incident beam, if the sample is thin enough, or deduced from the evolution of the intensity of XRF emission versus the energy of incident beam. Depending on the characteristics of beamlines, the available energy range around an atom absorption edge can be limited to the near-edge energy range and thus only the X-ray absorption near-edge structure (XANES) spectrum is accessible or on the contrary, large enough to allow the full recording of extended X-ray absorption fine structure (EXAFS) spectrum. In theory, nothing prevents to acquire step by step all the XANES spectra of a 2D mapping with the exception of the recording time. One square millimeter mapping with a pixel of one square micrometer requires the recording of one million spectra. The acquisition of one spectrum requiring a few minutes with standard detection systems and several days would be necessary to record such a map. Except with ultrafast detectors [6], the reconstruction of pixel XANES spectra from the scanning of area at the different energy is not faster. Actually, in addition to the element maps, only the valence state maps of elements can be obtained since it requires scanning the sample area only for a few number of energy around one absorption edge of the selected element. The recording of full XANES spectra is uniquely reserved for a few points of the map. To obtain the XANES spectra of a square millimeter area with a submicrometer resolution, that is, a few millions of XANES spectra with reasonable acquisition times, an approach of full-field type has been developed that we are now going to describe.

2.1 Full-field XANES (FF-XANES) analysis with nonfocused beam

Instead of scanning the interest area with a focused micro (or submicro)-X-ray beam, the interest area is directly illuminated using a larger beam, which can be obtained in removing all focusing optics as this is performed on ID21 (lensless setup) [2]. On this beamline, the sample X-ray radiography is recorded with a detection system consisted of a scintillator and a CMOS camera coupled to a long working distance optical objective in order to record a magnified transmission image of the sample. Two objectives with different magnification (10× or 20×) can be used, leading to a pixel size of 0.65 × 0.65 μm^2 or 0.32 × 0.32 μm^2 and a field of view of 1.5 × 1.5 mm^2 or 0.75 × 0.75 mm^2, respectively, for the 10× and 20× magnifications.

The 2D-XANES map of a selected element is obtained by recording a series of a few hundred X-ray radiographies at different energies, the energy being tuned in small steps (0.2–10 eV) across the absorption edge of selected element. Then a data preprocessing, including image alignment, normalization, and several other corrections, allows for creating the full-field XANES data set, which represents a 3D radiography stack consisting of a few hundred images recorded at different energies. Therefore, each pixel contains a full high-resolution XANES spectrum [2, 7].

2.2 Full-field XANES in transmission X-ray microscopy

The advent of high-performance X-ray sources and the improvement in the X-ray lens objectives allowed the development of transmission X-ray microscopy (TXM) using hard X-rays with spatial resolutions down to 30–40 nm and 15–30 μm FOV [7–9]. As previously, this type of device can be used to record a series of radiographs through the absorption edge of an element of interest. Besides the difference of spatial resolution and FOV, the data preprocessing is little more different because of the more complex design involving several X-ray lens whose focal lengths are energy dependent. In fact, the data processing is nontrivial and it is only with the design of new algorithms at the beginning of this decade that it was possible to obtain usable chemical images from TXM [7]. Currently, these algorithms are well integrated in most preprocessing software available on the TXM beamlines, or in data processing software such as TXM-Wizard package. Like the pixel size and the FOV are energy dependent, the spatial resolution and the FOV of 2D XANES image are, however, limited to the smallest FOV and the largest pixel size of image series. The concept of XANES microcopy is illustrated in **Figure 1** of Ref. [7].

Most of the software allow us to combine several image series in order to obtain a larger 2D XANES image and thus to investigate an area larger than the FOV of TXM used. However, this mode (called mosaic mode) can be rapidly time consuming and for investigating large areas, it can be more efficient to scarify the resolution and to use the nonfocused geometry outlined in the previous paragraph (Section 2.1).

2.3 Data processing

A 2D XANES image contains several millions of XANES spectra, and it was necessary to develop specific collections of programs to process these huge data, which were integrated in existing software package as, for instance, in PyMca package [10] or gather in new software such as in TXM-Wizard [11]. The latter was used for data processing of examples described in the next paragraphs. The reference described in detail the main features of toolkits available within the package. This software can be downloaded free from its web page https://sourceforge.net/projects/txm-wizard/ as well as reference papers.

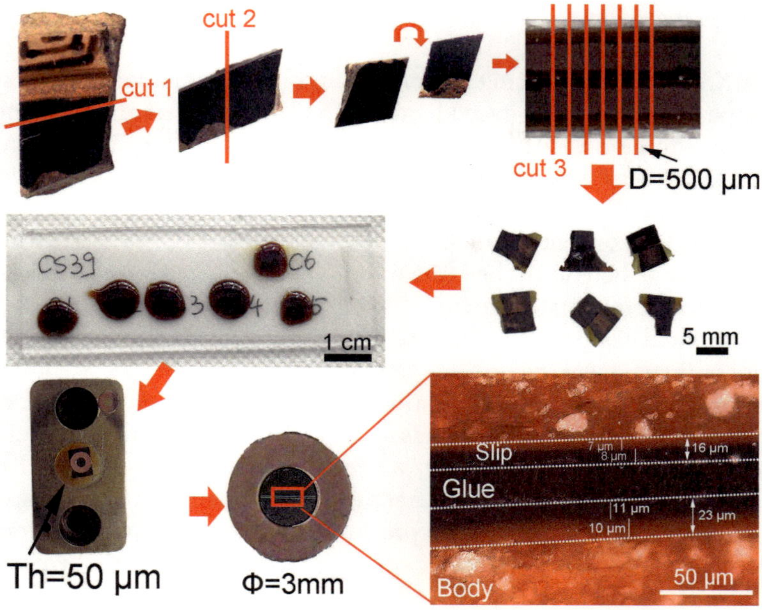

Figure 1.
Main steps in the cross-section sample preparation process for full-field XANES analysis. From left to right: (1–2) cutting using a wire saw in order to obtain (3) two strips with black decors, (4) sandwich making using glue, (5) cutting thin lamella (500 µm), (6) protection with glue, (7) mechanical polishing down to 40–50 µm, (8) adding a reinforcing Cu washer, and (9) finally optical micrograph of central area.

3. Cultural heritage materials studies: some examples in ancient ceramic field

To illustrate the power of FF-XANES analysis in the cultural heritage field, we chose to focus our purpose on ceramic material, one of the first synthetic materials made by human communities. Ceramic artifacts are also one of the most studied objects by the archeologist community [3]. Often produced in large amount by quite controlled manufacturing processes, this material is perfectly well suitable for in-depth studies inquiring sampling and specific sample preparations. Mass production gives easily access to material, and the control of manufacturing processes allows for limited investigations to a small corpus.

3.1 Iron speciation in Greek and Roman potteries

The first example concerns the Attic and Campanian potteries produced during the Greek and Roman periods, respectively. These two types of tableware are characterized of a red body partially or totally covered with a thin black high-gloss coating. Red body and black coating were obtained both from two different Fe-rich clay preparations and fired together using a complex firing protocol (lost now) allowing to obtain Fe^{2+}-based crystals in coating and Fe^{3+}-based crystals in body. Since the late eighteenth century, diverse protocols have been proposed, and even if the firing protocol in three steps (oxidative°➔°reductive°➔°oxida tive) theorized by Noble in 1965 is commonly accepted, the understanding of this technology is still too limited, especially to follow evolutions or adaptation to the local raw clays. As many fragments of these ceramics have survived to the present day, the use of investigation techniques requiring the preparation of suitable samples is not a problem but a good way to recover information on this lost technology [12].

3.1.1 Sample preparation

The FF-XANES analysis being performed in transmission mode, the best sample shape is a thin parallel-faced lamella. The lamella orientation depends on the sought information and its thickness can be estimated taking into account the concentration of the selected element and the average elemental composition. The global composition allows for calculating the X-ray transmission before the absorption edge, while the concentration of the selected element allows us to estimate the absorption edge height, in the same way as commonly performed for XANES measurements in transmission mode.

In the presented examples that have been partially published in Refs. [13, 14], the element of interest was the iron and the K-edge was chosen. The orientation of blade was chosen for studying Fe speciation from the sample surface to the body. The average composition of the coatings leads to a calculated optimal thickness of about 20–30 μm, while the one of the body (lower Fe rate) is around 40–50 μm. The intensity of incident X-ray beam being higher at the center, the surface coating has been placed at the center of the blade using a preparation method derived from transmission electron microscopy. The main steps of sample preparation are shown in **Figure 1**.

3.1.2 FF-XANES analysis

The radiographies were recorded on ID21 at ESRF using the 20× objectives, and the data preprocessing allowing to create the FF-XANES stacks was performed with the in-house software package PyMca [10] using the image realignment described in Ref. [15] and now included in the PyMca package. The closer data analysis was carried out using the TXM-Wizard software following the process described in Ref. [11]. **Figure 2(a)** shows one of the radiographies recorded before the Fe K-edge (at 7070 eV). The glassy coating appears as a dense and homogeneous layer of around a few tens microns, while the body reveals a more heterogeneous and porous structure. The first information, which can be extracted without further data processing, are the edge height and the edge energy position of all pixels (**Figure 2**). Filters allow for removing pixels for which either the edge height is too small regarding the noise level or the spectrum cannot be normalized. These pixels, as the pixels of glue, correspond to the iron-free areas or in too weak Fe concentration areas to give a XANES spectrum. The edge height map, correlated to the iron concentration, shows pretty well the difference in concentration between the coating and the body. It gives the same information as an element map of Fe. The edge energy position is proportional to the valence state: low energy corresponds to Fe^{2+} and higher energy corresponds to Fe^{3+}. The difference between the coating and the body is obvious and in accordance with their respective color.

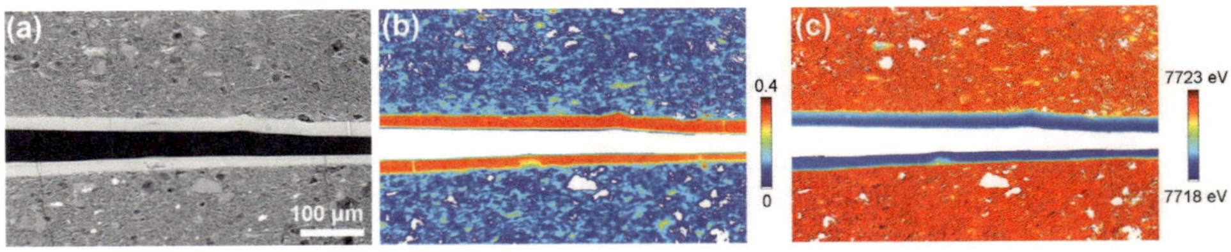

Figure 2.
First steps of data processing: (a) radiography image recorded at 7070 eV, that is, below the Fe-K absorption edge. The images were collected at ESRF beamline ID21 in transmission mode with a total FOV of 332.16 × 522.24 μm² (1038 × 1632 pixels). (b) Edge jump map linked to the Fe rate and (c) edge energy map linked to the iron state valence calculated from the XANES spectra evolution. The scale bar for (b) reports the values of the (absorption) edge jump, defined as the difference between the average intensity value in the XANES postedge region and the average intensity value in the XANES pre-edge region. The scale bar for the maps shown in (c) indicates the energy of the (absorption) edge in eV.

The study of the iron speciation needs a further analysis since it requires the use of the whole XANES spectrum and not only two parameters (height and energy position of jump edge) deduced from this spectrum. The number of pixel spectra is too huge (close to 2 millions) to consider an individual treatment. TXM-Wizard software proposes two approaches for a closer inspection of the set of XANES spectra: (i) a principle component analysis (PCA) followed by a *k*-means clustering or (ii) a least squares linear combination (LSLC) fitting with standards associated to a R-factors correlation analysis [11]. The first approach allows us to obtain a first result with open-minded solution. Here the PCA results show that the first four components (**Figure 3**) allow a rather good system description. Indeed, the first two components represent already a percentage of 99.6 and with the two others the value is about 99.8%. The *k*-means clustering, with k = 4, leads actually two different types of XANES spectra (**Figure 4a**). A clustering, with k > 4, does not give more different cluster types. The first cluster includes all the XANES spectra related to the coating, while the other clusters correspond to the body (**Figure 4b**). The XANES spectrum associated to cluster one shows some characteristics of hercynite ($FeAl_2O_4$) and magnetite (Fe_3O_4) reference spectra, while the three others are very close to maghemite (γ-Fe_2O_3) spectrum. The fitting of cluster 1 spectrum leads to 23% hercynite and 77% magnetite. The fitting of the two other cluster classes leads to similar results with 94–96% maghemite, 3–2% magnetite, and 3–2% hercynite. Even if hematite (α-Fe_2O_3) was not found in the clustering approach, this phase was considered in the LSLC fitting with standards for being sure that it is not present in some small areas, which was confirmed by the detection of no measurable amount of hematite (**Figure 5**). Maghemite is the main iron-based phase of body with a rate almost everywhere above 90%. In the coating, hercynite and magnetite share the Fe ions in an H/M ratio varying from 1 to 2. A few differences can be observed between

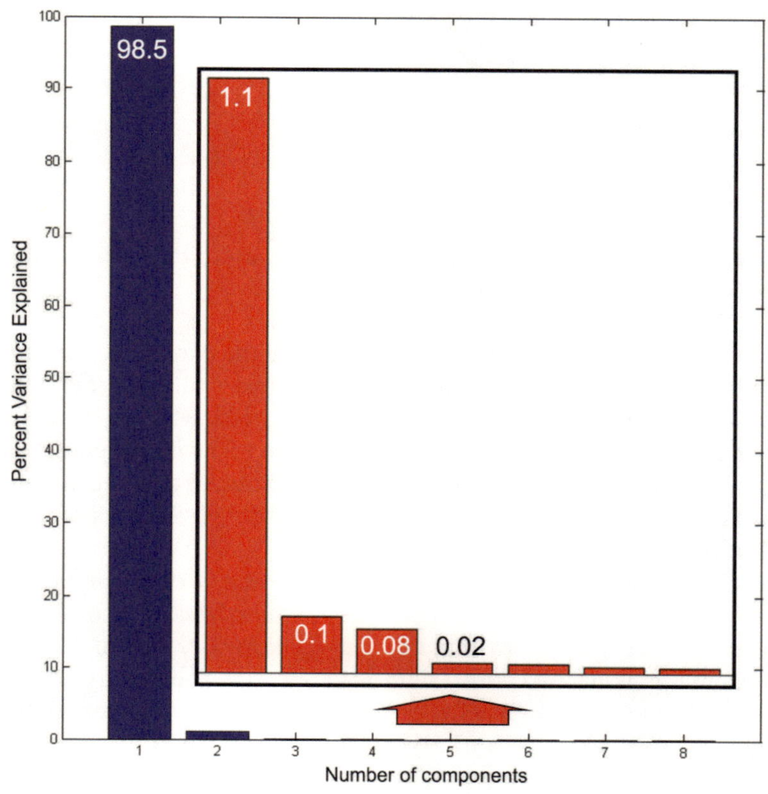

Figure 3.
Result of PCA analysis showing the contribution of the first principal components highlighted in blue color.

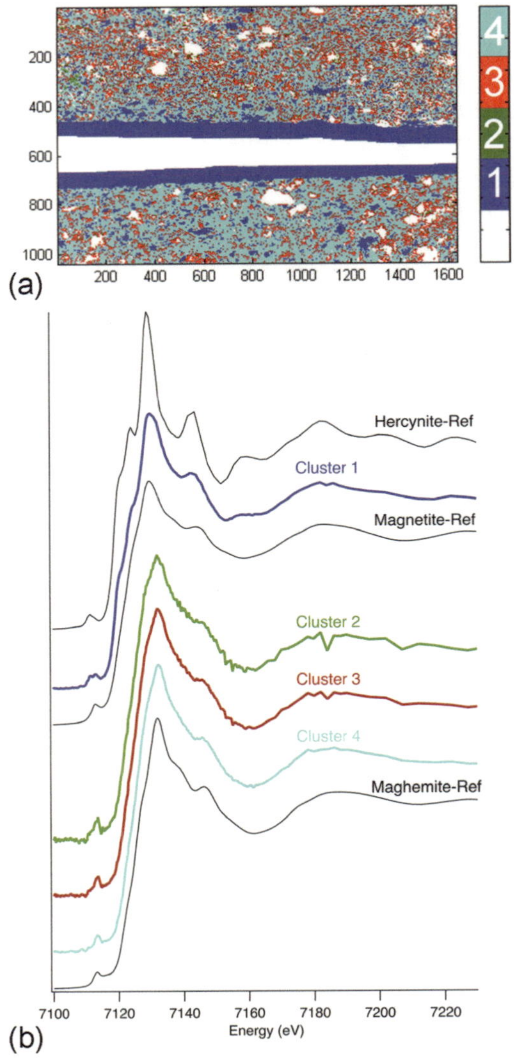

Figure 4.
(a) Result of clustering and (b) the associated spectra compared to spectrum references. The references of hercynite, magnetite, and maghemite are taken from ALS database.

the external (top) and internal (bottom) coatings. In the two cases, the hercynite concentration is higher in the upper part of coating (close to the surface), but the external coating exhibits a clear reoxidation of its surface, which is marked by the presence of a thin maghemite layer at its surface. The trichromatic phase map, excluding hematite, is shown in **Figure 6a** and compared to the map obtained for a Campanian ceramic fragment of Roman period. The detailed analysis for this last case can be found in Ref. [13].

The coating of Campanian contains mostly hercynite with only a few amount of maghemite mainly located at some surface areas, while the coating of Attic contains both hercynite and magnetite. In both cases, maghemite is the main Fe-based phase of the body but it is associated with a significant amount of hematite or hercynite in localized areas of Campanian body. These differences do not come from small variations in chemical composition between these two types of potteries, made at different places and times from similar but, however, different raw materials. They are actually due to the firing conditions. The iron valence repartition, mostly Fe^{2+} and Fe^{3+} in the coating and body, respectively, is consistent with the Noble's

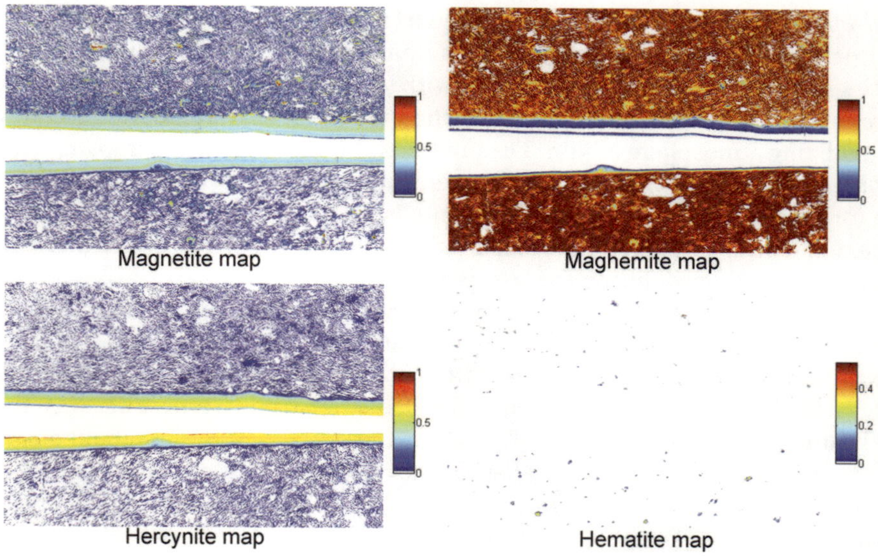

Figure 5.
Phase maps obtained from the least squares linear combination fit of reference XANES spectra for magnetite, maghemite, hercynite, and hematite for each single pixel.

protocol. During the reducing step (step 2), the highest temperature of the firing is achieved and a significant part of iron is Fe^{2+}; Fe^{2+}/Fe^{3+} ratio increases with the temperature. The coating has a lower glazing temperature than the body and thus is also glazed during this step. Consequently, much more limited chemical reactions occur between the coating and the kiln atmosphere compared to the body and the kiln atmosphere during the last oxidative step achieved at a lower temperature than in the step 2. Then, during this last step, the body is mainly oxidized and turns red. The higher rate of hercynite and the absence of magnetite in the glazed coating of Campanian indicate that the temperature during the step 2 was higher here than for Attic. On the other hand, the presence of maghemite in some areas of the coating indicates that the chemical reaction between the coating and the kiln atmosphere was stronger during the last step. Maghemite is not distributed in a thin surface layer as in the exterior Attic coating but it is distributed in depth in some localized areas. The amount in these areas is also higher. This feature seems to be due to worse glazing of the Campanian coating. Coatings are made from the fine part of raw clay, and the selected part for the elaboration of Attic coating was

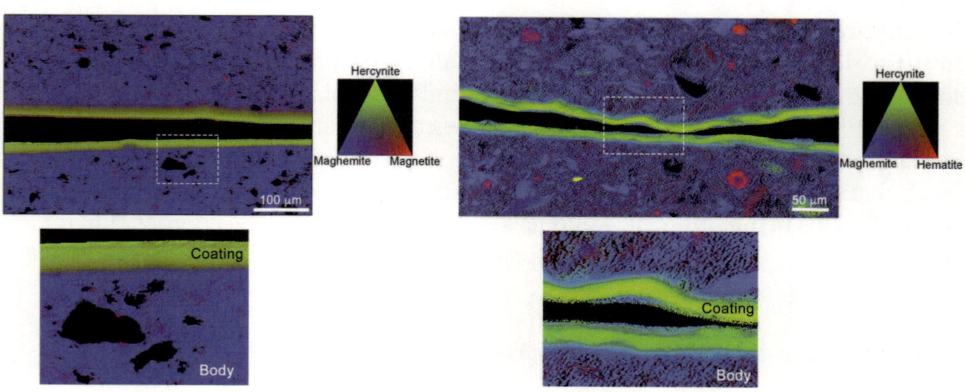

Figure 6.
Comparisons of the phase maps of Attic (left) and Campanian (right) ceramics [13, 14].

finer than for Campanian as revealed by quartz crystals sizes. Also, even if the two coatings have close elemental compositions, the glazing temperature of Attic coating is lower. The use of less fine part with larger grains of sand (mainly quartz) increases the porosity of glazing coating and then enhances its oxidation rate under the oxidizing atmosphere of kiln. The presence of hematite in the Campanian body is consistent with an oxidation at higher temperature allowing a partial transformation (or recrystallization) of maghemite into hematite. The hercynite-rich zones of Campanian body correspond to partially glazed zones, which did not be therefore completely reoxidized during the last step.

A thin blade prepared from the same Campanian fragment was also analyzed using the transmission X-ray microscope of 6-2c beamline at the Stanford Synchrotron Radiation Lightsource (SSRL), SLAC National Accelerator Laboratory [13]. The differences are only related to the characteristic differences between the two beamlines, that is, a smaller FOV and a better spatial resolution (about one order of magnitude) at Stanford. The FOV can be enlarged using the mosaic mode, which consists of recorded several series of radiographies with overlapping areas in order to obtain a 2D XANES map larger than the FOV. However, the investigated area cannot reach the FOV of ID21. In the problematic presented here, the better resolution does not bring much more in the iron speciation study and the larger FOV of ID21 shows a best advantage. On the other hand, the very good resolution of 6-2c beamline allowed for studying the sample porosity and for showing that hercynite was distributed in well-glazed dense zones. Maghemite is present in zones of greater porosity confirming that the hercynite is formed during the step 2 before the glazing. The maghemite comes from the oxidation of the last one, which is in contact with oxygen thanking to the open porosity during the step 3. A study of a presigillata fragment was also carried out at SSRL [13]. This type of pottery, which was also made during the Roman period but a century later than the Campanian productions, presents also a glazed black coating covering a red body. The results revealed both a high concentration of hercynite in the coating and a significant rate of hematite in the body. The top layers of the coating showed also a strong reoxidation with a significant presence of maghemite at the surface. The firing temperature during the step 2 was similar to the one of the analyzed Campanian but the oxidizing step 3 seems to start at higher temperature.

To conclude with this explanation of the problematic, a few words will be given on a recent study, which suggests that more complex firing protocols would have been used for some Attic productions [16]. The study was carried out on microsamplings coming from a decorated fragment attributed to the Berlin's painter and belonging to the collection of the J. Paul Getty Museum showing as well the actual possibility to use this technique to study precious samples. The analyzed fragment presents in some areas a red coating under the black coating ([16], **Figure 1**). FF-XANES investigation revealed that this red intermediate coating contained mainly hematite, while hercynite and maghemite were found in the black coating and the body, respectively ([16], **Figure 2**). The massive presence of hematite in this intermediate coating, whose chemical composition and the porosity are identical to the ones of black coating, was not considered as consistent with the firing protocol proposed by Noble. This firing protocol would have led to the same mineral composition for the two coatings. Also the authors proposed a more complex protocol involving two separate firings.

3.2 Under glaze decors of Chinese porcelain

In the first example, the average elemental composition and the concentration of the selected element, in this case iron, were quite suitable to transmission mode measurements. This technical analysis can be also successfully used to study the speciation of an element in low concentration with, however, some restrictions as described in this second example. This example concerns the speciation of the coloring elements of the famous blue and white porcelain produced during the Ming Dynasty (1368–1644 AD).

Blue and withe porcelain (**Figure** 7), so-called Qinghua porcelain, is characterized of a brilliant white translucent glaze with a blue pattern painted underneath. The color is due to cobalt, but other transition metal elements including iron and manganese are also present in the Ming productions [17] and the aim of study was for establishing the feature of each [18].

3.2.1 Sample preparation

The thin parallel-faced blades were made as previously described (cf. Section 3.1.1) expect that it was not possible to determine an optimal thickness from the average element composition. The average concentration in transition elements (Fe, Mn, and Co) is very low (≤1%) and necessitates a thick sample superior to 500 μm to have a right edge jump at the K-edge of cobalt, for instance. Unfortunately with such a thickness, the absorption of these other elements (Al, Si, K, and Ca) limits the transmission in the energy range to a few percent and does not allow measurements in transmission mode. Also to conserve a right transmission in the investigated energy range, the sample thickness was chosen between 40 and 60 μm [18]. A sample ready for investigation is shown in **Figure 8**.

Figure 7.
Qinghua porcelain, Mieping vase with "Eight dragons among the clouds." Ming Dynasty (1368–1644), Yongle reign (1403–1424), Jingdezhen workshop (Jiangxi province, China). Provenance: Estate of Jutta Frieda Luise Meischner and Bolurfrushan Family, Inv.-Nr. MBS 06/16–6.9.02 (© Æli Barjesteh, ASET Stiftung).

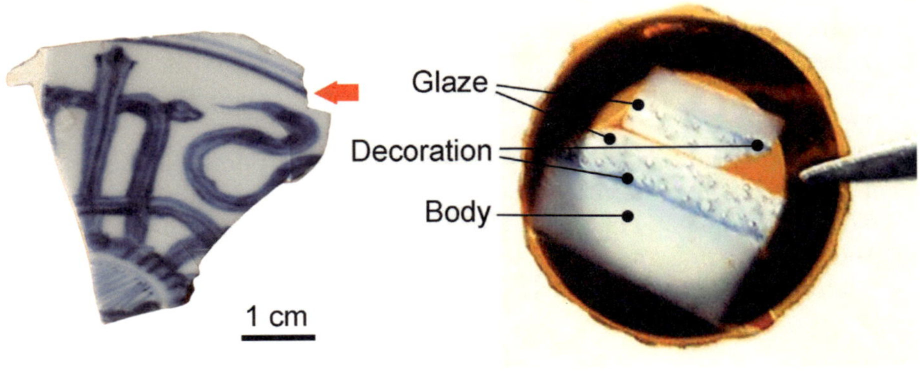

Figure 8.
Sample for FF-XANES analysis (right) obtained from a fragment of Ming blue and white porcelain (left). For the preparation method, refer to **Figure 1**.

3.2.2 Investigations

X-ray absorption spectra (XAS) collected in fluorescence mode are better suited for determining the speciation of elements in low concentration [19]. Unfortunately, the scanning of a large zone with a submicrometric resolution is too time-consuming for analyzing even a few samples. Also the technical advantages of the two techniques were combined.

At first, XRF maps were collected in scanning mode on large areas. The PyMca software was used to batch-fit the XRF spectra, to generate the elemental maps (from Na to Co) and to normalize these maps taking into account the incident beam intensity [10]. A typical result showing the main differences of elemental distribution in the various layers of a Ming Qinghua porcelain is presented in **Figure 9**. The colored zone is clearly identified by the high rate of calcium. The localized high concentrations in silicon correspond to quartz grains, which are more numerous in the body than in the glaze. The needle-shape crystals observed in the colored zone containing calcium are anorthite crystals [20]. The distribution of three transition elements (Mn, Fe, Co) is quite different. Mn and Co distributions are concentrated in colored areas confirming there are well contained into the pigment. However, Mn distribution is more uniform and does not present local high level compared to Co element. Fe is quite homogeneously distributed both in colored area and in above glaze. Laboratory XRF measurements revealed that the glaze areas without blue decors contain similar Fe rates [18]. A few punctual XANES spectra were acquired at the Co K-edge both over Co-rich particles and diluted Co in glassy matrix, by scanning the energy of the incoming beam and measuring the X-ray fluorescence signal of Co K-lines. For Co-rich particles, the recorded spectra were similar to the one of cobalt aluminate crystals ($CoAl_2O_4$), while in lower concentrated zones, the recorded spectra correspond to Co^{2+} diluted in glass ([18], **Figure 5**). These spectra were acquired only for a few points. In order to have a full 2D picture of Co speciation and in particular to verify that all Co-rich particles are cobalt aluminate and not cobalt oxide phase, FF-XANES maps were acquired in a second step.

Series of radiographies were recorded through the Co K-edge, and the data processing was carried out as previously described. **Figure 10** shows one of the obtained results. The edge energy position map confirmed the presence of the only one state of valence and the phase map, the two types of cobalt speciation, that is, Co^{2+} in the glassy matrix and Co^{2+} in cobalt aluminate crystals. This last map was obtained by LSLC fitting after a PCA analysis and *k*-means clustering revealing the existence of only two types of XANES spectra. However, for the majority of the

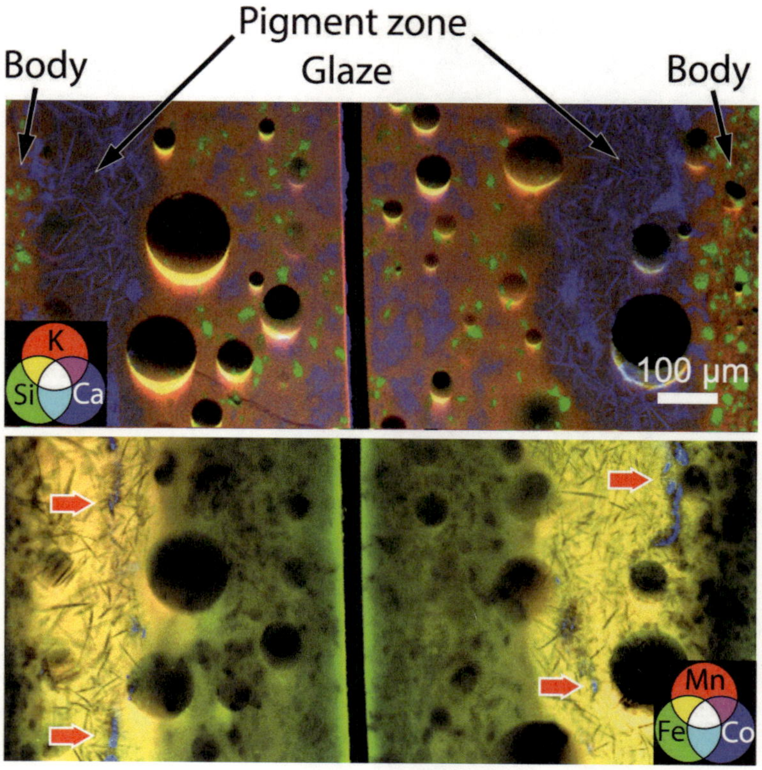

Figure 9.
Tricolor elemental maps obtained from μSR-XRF measurements performed on ID21 beamline at ESRF [18].

investigated zones, only Co-rich particles were studied. The cobalt concentration outside the particles was too low to give worktable pixel XANES spectra. For two samples, the Fe level was high enough to allow FF-XANES investigations at the Fe K-edge. The results confirmed the absence of Fe-based particles and a rather homogeneous distribution around the long narrow strips (anorthite crystals) and the cobalt aluminate particles. No variance of the edge energy was observed and the PCA confirmed the existence of only one type of XANES spectrum close to the one of bivalent iron dispersed in a glassy matrix [18]. Micro-X-ray diffraction is also available on ID21 beamline and was used to identify the phase structure of crystals. These measurements have also revealed that Co is partially substituted by another metal element in cobalt aluminate crystals.

To conclude this first part, X-ray scanning microscopy associated to FF-XANES analysis is an efficient way to determine the complex structure of underglaze decors

Figure 10.
FF-XANES investigation performed on ID21 beamline at ESRF [18]. (a) Transmission image recorded at 7670 eV, (b) edge energy map, (c) phase map obtained from the least squares linear combination fitting (standards $CoAl_2O_4$ and Co in glaze) of each single pixel.

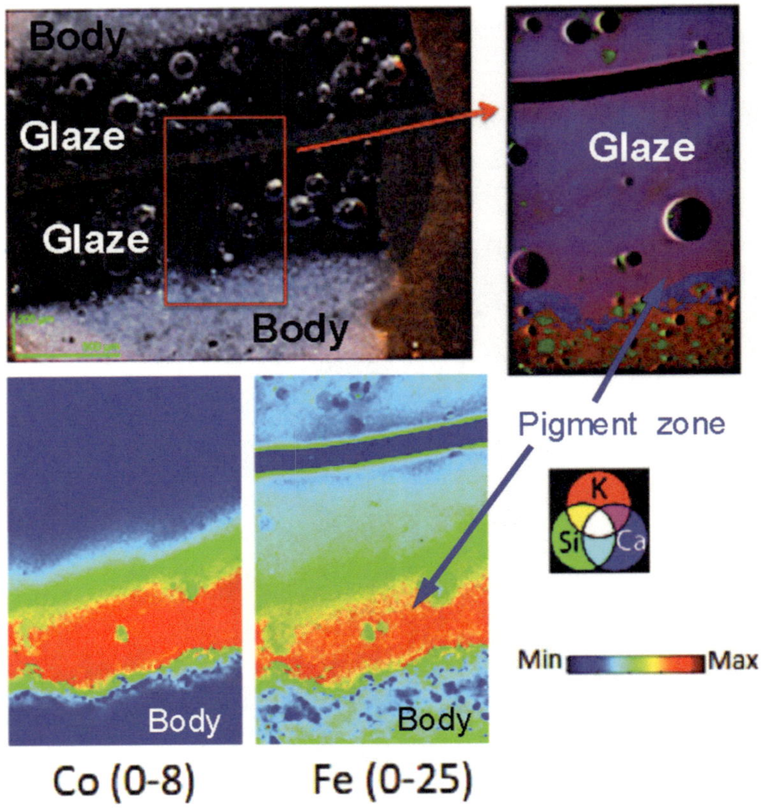

Figure 11.
Typical elemental maps of a Yuan sample obtained from μSR-XRF measurements performed on ID21 beamline at ESRF. The transition elements (Co, Fe) are concentrated inside the pigment zone.

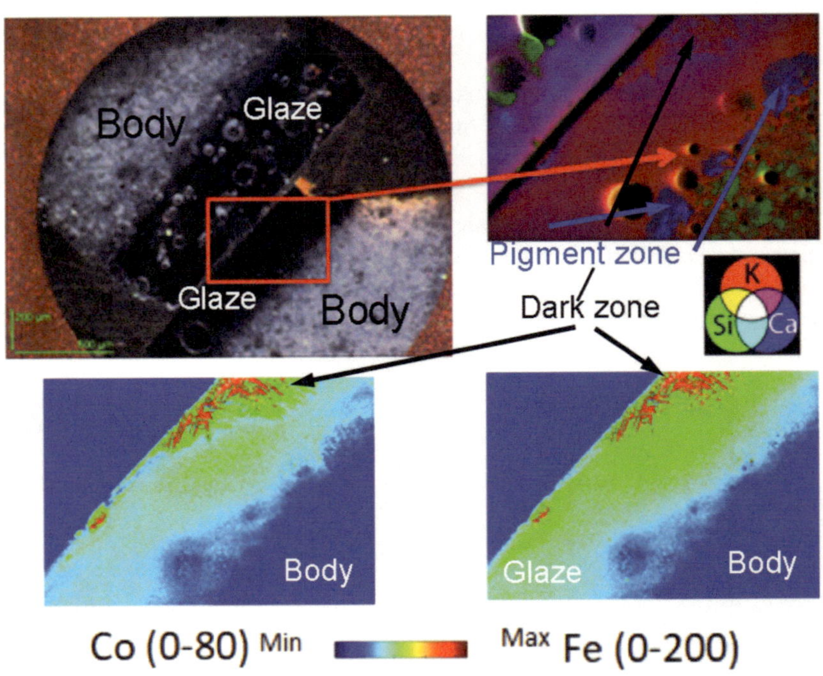

Figure 12.
Elemental maps obtained from a dark zone showing the $CoFe_2O_4$ dendritic crystals.

of Chinese ceramics. It allows not only to identify the color pigments and the speciation of colorant ions but also to observe how they are distributed in the glaze matrix.

These interesting first results led us to analyze previous productions made during the Yuan Dynasty (1269–1378 AD), which is considered by historians as the heyday of this type of porcelain. The elemental compositions of the two types of productions are quite close, except for Yuan blue color, which contains no Mn and much more Fe as confirmed by the elemental maps deduced from the X-ray fluorescence emission collected in scanning mode. The spatial distributions of the main elements (Al, Si, K, and Ca) are similar to the ones of Ming samples. The main difference concerns only the distributions of transition elements (Fe and Co). A higher concentration of Co clearly appears in the colored area close to the body/glaze interface, but the distribution is quite homogeneous unlike for Ming samples (**Figure 11**). No Co-rich particle was found in this area. The Co distribution is only diffuse and similar to the one of Fe element. In some cases, high punctual Co contents were only observed at the top surface and always associated to high Fe content as shown in **Figure 12**. In fact, these specific zones correspond to the dark spots visible at the surface of blue decorations. This situation did not allow for performing FF-XANES investigations at Co K-edge as previously done. The absorption before the Co K-edge is too strong because of the significant iron absorption. A reduction of the sample thickness would decrease the Fe absorption but would also decrease the Co absorption, making it unmeasurable. Consequently, FF-XANES investigations have been possible only at the Fe K-edge.

4. Conclusion

FF-XANES analysis is a very powerful technique to study complex structure and provide pertinent chemical information. It combines spatial and energy resolution with large FOVs and fast acquisition in a virtually limitless variety of samples. The spatial resolution can be adapted to the problematic from a few hundred to a few tens of nanometer following the geometry used (defocalized beam or transmission X-ray microscope). Software packages, such as TXM-Wizard or PyMca, provide user-friendly tools to successfully plan and collect the series of radiographies, to build from these images the 2D XANES map (advanced averaging, image alignment, filtering), and to process these big data in order to extract chemical information such as rate, valence state, and speciation. These tools are rather easy to use and allow obtaining pertinent information in the form of easily exploitable maps.

This technique combines XANES analysis (selectivity) and mapping of large areas at high revolution advantages. For complex systems consisting of many elements, the possibility to focus on specific key element is fundamental, while for heterogeneous system, the possibility to investigate a large area with a high spatial resolution is essential. FF-XANES analysis is thus ideally suited for studying complex and heterogeneous materials such as materials involved in the conception of ancient objects. In particular, the relationships between a specific physical property and the involved elements are quickly determined. For instance, in the second example presented in previous paragraphs, the distribution and the speciation of transition elements to the color are linked. In the first example of reverse engineering, iron is the key element to identify the firing process and the selectivity of technique allows focusing on the Fe-based phases without being disturbed by the other phases. This technique begins to be used with success on other cultural heritage material types such as pigment in paintings of Henri Matisse [21].

This analytical technique is based on measurements in transmission mode, which implies a specific preparation of samples and the limitations of the collecting XANES spectra as illustrated in the first example.

Of course, this interesting technique is not reserved to cultural heritage materials, and several other examples of study can be found in the literature concerning different material types [22–26]. Full-field-based techniques are still in development and become more and more available on synchrotron beamlines such as, for instance on PUMA, a beamline at SOLEIL (French synchrotron, Saclay) dedicated to the study of cultural heritage materials.

Acknowledgements

The authors would like to thank Æli Barjesteh (ASET Stiftung) who kindly provided the photography of the Ming vessel. We would like also to thank Philippe Goudeau (Pprime Institut, Poitiers) and Marine Cotte (ESRF, Grenoble) for their suggestions and comments. Research was supported by the CAI YUANPEI 2016–2018 program (No. 36708RD), the preresearch foundation of Shaanxi University of Science and Technology (No. 2018GBJ-08), and the ARCHIMEDE Labex program: Investissement d'Avenir ANR-11-LABX-0032-01.

Author details

Philippe Sciau[1*] and Tian Wang[2*]

1 CEMES-CNRS, Toulouse University, Toulouse, France

2 School of Materials Science and Engineering, Shaanxi University of Science and Technology, Xi'an, PR China

*Address all correspondence to: philippe.sciau@cemes.fr and wangtian@sust.edu.cn

References

[1] Bertrand L, Cotte M, Stampanoni M, Thoury M, Marone F, Schöder S. Development and trends in synchrotron studies of ancient and historical materials. Development and Trends in Synchrotron Studies of Ancient and Historical Materials. 2012;**519**:51-96. DOI: 10.1016/j.physrep.2012.03.003

[2] Cotte M, Pouyet E, Salome M, Rivard C, De Nolf W, Castillo-Michel H, et al. The ID21 X-ray and infrared microscopy beamline at the ESRF: Status and recent applications to artistic materials. Journal of Analytical Atomic Spectrometry. 2017;**32**:477-493. DOI: 10.1039/c6ja00356g

[3] Sciau P, Goudeau P. Ceramics in art and archaeology: A review of the materials science aspects. European Physical Journal B. 2015;**88**:132. DOI: 10.1140/epjb/e2015-60253-8

[4] Cotte M, Genty-Vincent A, Janssens K, Susini J. Applications of synchrotron X-ray nano-probes in the field of cultural heritage. Comptes Rendus Physique. 2018;**19**:578-588. DOI: 10.1016/j.crhy.2018.07.002

[5] Dejoie C, Tamura N, Kunz M, Goudeau P, Sciau P. Complementary use of monochromatic and white-beam X-ray micro-diffraction for the investigation of ancient materials. Journal of Applied Crystallography. 2015;**48**:1522-1533. DOI: 10.1107/S1600576715014983

[6] Monico L, Janssens K, Alfeld M, Cotte M, Vanmeert F, Ryan CG, et al. Full spectral XANES imaging using the Maia detector array as a new tool for the study of the alteration process of chrome yellow pigments in paintings by Vincent van Gogh. Journal of Analytical Atomic Spectrometry. 2015;**30**:613-626. DOI: 10.1039/C4JA00419A

[7] Meirer F, Cabana J, Liu Y, Mehta A, Andrews JC, Pianetta P. Three-dimensional imaging of chemical phase transformations at the nanoscale with full-field transmission X-ray microscopy. Journal of Synchrotron Radiation. 2011;**18**:773-781. DOI: 10.1107/S0909049511019364

[8] Chu YS, Yi JM, De Carlo F, Shen Q, Lee W-K, Wu HJ, et al. Hard-X-ray microscopy with Fresnel zone plates reaches 40nm Rayleigh resolution. Applied Physics Letters. 2008;**92**:103119. DOI: 10.1063/1.2857476

[9] Liu Y, Andrews JC, Wang J, Meirer F, Zhu P, Wu Z, et al. Phase retrieval using polychromatic illumination for transmission X-ray microscopy. Optics Express. 2011;**19**:540-545. DOI: 10.1364/OE.19.000540

[10] Solé VA, Papillon E, Cotte M, Walter P, Susini J. A multiplatform code for the analysis of energy-dispersive X-ray fluorescence spectra. Spectrochimica Acta Part B: Atomic Spectroscopy. 2007;**62**:63-68. DOI: 10.1016/j.sab.2006.12.002

[11] Liu Y, Meirer F, Williams PA, Wang J, Andrews JC, Pianetta P. TXM-wizard: A program for advanced data collection and evaluation in full-field transmission X-ray microscopy. Journal of Synchrotron Radiation. 2012;**19**:281-287. DOI: 10.1107/S0909049511049144

[12] Sciau P, Leon Y, Goudeau P, Fakra SC, Webb S, Mehta A. Reverse engineering the ancient ceramic technology based on X-ray fluorescence spectromicroscopy. Journal of Analytical Atomic Spectrometry. 2011;**26**:969-976. DOI: 10.1039/c0ja00212g

[13] Meirer F, Liu Y, Pouyet E, Fayard B, Cotte M, Sanchez C, et al. Full-field XANES analysis of Roman ceramics

to estimate firing conditions—A novel probe to study hierarchical heterogeneous materials. Journal of Analytical Atomic Spectrometry. 2013;**28**:1870-1883. DOI: 10.1039/c3ja50226k

[14] Wang T. A multi-scale study of ancient ceramics using a series of analytical techniques [PhD thesis]. Toulouse University; 2016

[15] Paleo P, Pouyet E, Kieffer J. Image stack alignment in full-field X-ray absorption spectroscopy using SIFT_PyOCL. Journal of Synchrotron Radiation. 2014;**21**:456-461. DOI: 10.1107/S160057751400023X

[16] Cianchetta I, Trentelman K, Maish J, Saunders D, Foran B, Walton M, et al. Evidence for an unorthodox firing sequence employed by the Berlin painter: Deciphering ancient ceramic firing conditions through high-resolution material characterization and replication. Journal of Analytical Atomic Spectrometry. 2015;**30**:666-676. DOI: 10.1039/c4ja00376d

[17] Wen R, Wang CS, Mao ZW, Huang YY, Pollard AM. The chemical composition of blue pigment on Chinese blue-and-white porcelain of the Yuan and Ming dynasties (AD 1271-1644). Archaeometry. 2007;**49**:101-115. DOI: 10.1111/j.1475-4754.2007.00290.x

[18] Wang T, Zhu TQ, Feng ZY, Fayard B, Pouyet E, Cotte M, et al. Synchrotron radiation-based multi-analytical approach for studying underglaze color: The microstructure of Chinese Qinghua blue decors (Ming dynasty). Analytica Chimica Acta. 2016;**928**:20-31. DOI: 10.1016/j.aca.2016.04.053

[19] Calvin S. XAFS for Everyone. Boca Raton: CRC Press; 2013

[20] Qu Y, Xu J, Xi X, Huang C, Yang J. Microstructure characteristics of blue-and-white porcelain from the folk kiln of Ming and Qing dynasties. Ceramics International. 2014;**40**:8783-8790. DOI: 10.1016/j.ceramint.2014.01.100

[21] Pouyet E, Cotte M, Fayard B, Salomé M, Meirer F, Mehta A, et al. 2D X-ray and FTIR micro-analysis of the degradation of cadmium yellow pigment in paintings of Henri Matisse. Applied Physics A: Materials Science & Processing. 2015;**121**:967-980. DOI: 10.1007/s00339-015-9239-4

[22] Harris WM, Nelson GJ, Izzo JR, Grew KN, Chiu WKS, Chu YS, et al. Full Field Imaging of Nickel Oxidation States in Solid Oxide Fuel Cell Anode Materials by XANES Nanotomography. New York: Amer Soc Mechanical Engineers; 2012

[23] Boesenberg U, Meirer F, Liu Y, Shukla AK, Dell'Anna R, Tyliszczak T, et al. Mesoscale phase distribution in single particles of $LiFePO_4$ following lithium deintercalation. Chemistry of Materials. 2013;**25**:1664-1672. DOI: 10.1021/cm400106k

[24] Yang F, Liu Y, Martha SK, Wu Z, Andrews JC, Ice GE, et al. Nanoscale morphological and chemical changes of high voltage lithium manganese rich NMC composite cathodes with cycling. Nano Letters. 2014;**14**:4334-4341. DOI: 10.1021/nl502090z

[25] Hesse B, Salome M, Castillo-Michel H, Cotte M, Fayard B, Sahle CJ, et al. Full-field calcium K-edge X-ray absorption near-edge structure spectroscopy on cortical bone at the micron-scale: Polarization effects reveal mineral orientation. Analytical Chemistry. 2016;**88**:3826-3835. DOI: 10.1021/acs.analchem.5b04898

[26] Sorbadere F, Laurenz V, Frost DJ, Wenz M, Rosenthal A, McCammon C, et al. The behaviour of ferric iron during partial melting of peridotite. Geochimica et Cosmochimica Acta. 2018;**239**:235-254. DOI: 10.1016/j.gca.2018.07.019

Chapter 4

Applications of Synchrotron-Source IR Spectroscopy for the Investigation of Insect Wings

Samuel Cheeseman, Vi Khanh Truong, Jitraporn Vongsvivut, Mark J. Tobin, Russell Crawford and Elena P. Ivanova

Abstract

Synchrotron-source infrared (IR) spectroscopy offers an effective method to characterise the chemical composition across surfaces. The intense light source allows the detection of trace quantities of different chemical components with a superior signal-to-noise ratio, while the highly collimated light enables high-resolution spatial mapping of the chemical distribution. In this chapter, we introduce synchrotron-source IR spectroscopy, using the infrared microspectroscopy (IRM) beamline at the Australian Synchrotron as an example. We then discuss the use of synchrotron-source IR spectroscopy to analyse insect wings in terms of experimental setup and a summary of the results in two different modes of operation, transmission and attenuated total reflection (ATR). Insect wings possess unique anti-wetting, self-cleaning, anti-biofouling and bactericidal properties and provide inspiration for biomimetic surfaces on synthetic materials which possess similar properties, useful in a range of industries.

Keywords: synchrotron, infrared, insect wings, IR spectroscopy, FTIR

1. Introduction

Infrared (IR) spectroscopy can be used for the chemical characterisation and identification of materials. The use of a synchrotron light source, which is extremely intense and highly collimated, enables a superior signal-to-noise ratio and high-resolution spatial mapping of materials and surfaces compared to alternative IR sources.

IR spectroscopy relies on the absorption of specific wavelengths within the IR spectra, upon interacting with certain molecular bonding. Differences in electronegativity and/or orientations of molecular species create a disproportional distribution of charges. Atomic elements that are more electronegative more strongly attract electrons when bonding with an element with lower electronegativity, resulting in dipole moments. As the molecules naturally vibrate, the distance between the negative charge centre of each elemental species in a specific bond fluctuates, creating an electric field. This fluctuation is known as the resonant frequency. If the frequency of the IR radiation, which changes according to wavelength, matches the molecular resonant frequency, the radiation is absorbed. The size of the molecular frequency is determined by the size and electronegativity of the atoms involved in the chemical bond, as well as the immediate molecular environment. These modes of vibration can include bending, scissoring, rocking and symmetric and asymmetric stretching [1]. Through detection of absorbed wavelength ranges, which typically form easily identifiable peaks, it is possible to determine the major chemical bonding present in the sample, indicative of the chemical composition [2, 3].

This chapter discusses the setup, advantages and results of synchrotron-source IR spectroscopy for the investigation of insect wings, with a particular focus on the setup at the Australian Synchrotron. The wings of particular insects are of interest as they possess unique anti-wetting, anti-biofouling and mechano-bactericidal properties due to their chemical composition and nanostructured surface [4–8]. These properties are useful as biomimetic templates for the design of anti-wetting and anti-biofouling materials, with applications in a range of industries such as the hulls of ships or water pipes [9]. Additionally, the mechano-bactericidal activity of these surfaces provide a novel solution to preventing bacterial colonisation on biomedical devices and other surfaces, particularly in light of emerging concerns of antibiotic resistance. These surfaces are advantageous to present strategies which rely largely on coatings embedded with antimicrobial agents resulting in issues of reduced efficacy over time and cytotoxic side effects and can promote the development of resistance [10–13].

2. Infrared beamline at the Australian Synchrotron

Infrared (IR) microspectroscopy, as the name suggests, allows the chemical composition of materials to be determined at the microscopic scale. Typically, an IR transmission microscope, with all reflecting optics, is coupled to a Fourier-transform IR (FTIR) spectrometer, permitting mid-IR absorption spectra to be collected from features as small as 5–10 μm in size within a sample to be analysed or identified. This process relies on the use of a high numerical aperture (NA) condenser and objective optics to focus the IR beam onto the sample, with the subsequent collection of the beam that has passed through the sample. The limit of the sample dimensions that can be examined is set by the optical diffraction limit at the analysing wavelength (e.g. 1600 cm^{-1} = 6.25 μm in wavelength). Selection of the region of interest for analysis is typically determined by inserting an aperture into the optical path of the IR beam at a focus position that is placed either before or after the sample. This results in a loss of most of the intensity of the beam from a conventional IR source, such as that of a Globar™ source, from reaching the detector. Alternatively, multi-pixel IR imaging detectors, such as linear array and focal plane array (FPA) detectors, can be used to enable the simultaneous collection of FTIR spectra from many positions within the sample. This approach, however, also typically relies on less than 0.1% of the total intensity of the IR source being directed to each sensing element of the FPA detector.

An alternative approach, described here, combines the high brightness of a broadband synchrotron source with an FTIR spectrometer and IR microscope, which enables the full intensity of the IR source to be directed to a diffraction-limited position on the sample. This results in a high signal intensity and a high signal-to-noise ratio (SNR) when analysing samples with features of 5–10 μm in size [14–16]. Synchrotron IR beamlines dedicated to the high spatial resolution analysis of microscopic materials are currently in operation at a number of national laboratories worldwide.

The Australian Synchrotron commenced operations for researchers in 2007, with two IR beamlines having been included in the initial suite of beamlines. One of these was dedicated to high spectral resolution gas-phase spectroscopy in the far-IR spectral region, while the other beamline was established for mid-IR microspectroscopy. Extraction of IR radiation from the Australian Synchrotron is achieved though the incorporation of a removable plane mirror (M1, see **Figure 1**) within the vacuum chamber of one of the dipole bending magnets of the electron storage ring. **Figure 1A** shows the modified vacuum chamber of the electron storage ring, with two ports incorporated to enable the insertion of the IR M1 mirror and extraction of the IR light reflected off the M1 mirror. **Figure 1B** shows the M1 mirror undergoing vibration testing prior to installation in the synchrotron storage ring. When inserted, the M1 mirror collects an angular portion of the emitted IR and visible synchrotron radiation of 58 mrad (horizontal) and 17 mrad (vertical), with this beam being reflected towards a toroidal focusing mirror (M2). This mirror focuses the beam downwards towards two plane mirrors M3 and M3a, which, in turn, direct the beam to a focus point through the shield wall at a position just beyond a chemical vapour deposition (CVD) diamond window that separates the vacuum of the synchrotron storage ring from the vacuum of the IR beamlines.

A schematic depiction of the beamline layout is given in **Figure 1C**. Beyond the CVD diamond window, the synchrotron beam expands and is directed to the beam splitter optics, where half of the horizontal fan of radiation is reflected to each of the two IR beamlines. The final matching optics for each of the IR beamlines is designed to provide a well-collimated beam that is matched in dimensions to the internal optics of the FTIR instrument, which is coupled to the beamline. In the case of the IR microspectroscopy (IRM) beamline, a combination of plane mirrors and a spherical focusing mirror delivers a collimated beam of approximately 12 mm × 12 mm in an area at the entrance port of the FTIR spectrometer. While this beam does not fill the interferometer beam splitter, it provides a beam that is matched in size to the entrance aperture of the reflecting condenser optic used for most experiments.

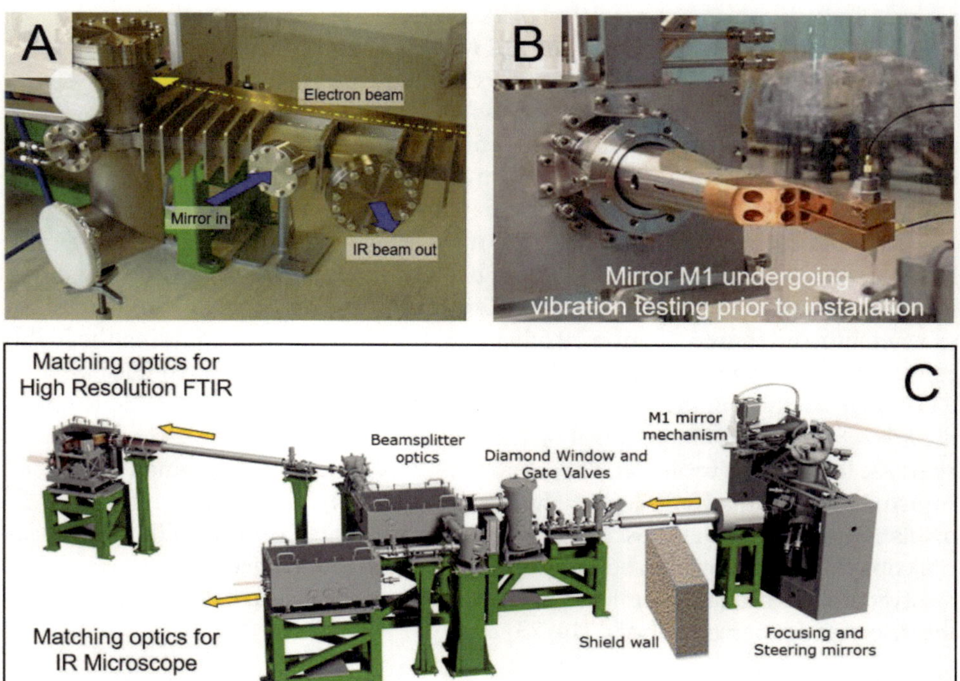

Figure 1.
(A) Modified vacuum chamber of the storage ring dipole magnet, with ports to allow the insertion of the M1 beamline mirror and the collection of the reflected radiation. (B) M1 mirror undergoing vibration testing. (C) Layout of the beamline optics of the IRM and THz/far-IR beamlines and their shared beamline front end.

3. Experimental setup for the investigation of insect wings

The IRM beamline at Australian Synchrotron is equipped with a Bruker VERTEX 80v FTIR spectrometer coupled with a Hyperion 2000 FTIR microscope and a liquid nitrogen-cooled narrowband mercury cadmium telluride (MCT) detector (Bruker Optik GmbH, Ettlingen, Germany). The synchrotron FTIR mapping measurements described in this chapter were performed within a spectral range of 3800–700 cm^{-1} using a 4 or 8 cm^{-1} spectral resolution. Blackman-Harris 3-Term apodisation, Mertz phase correction and zero-filling factor of 2 were set as default acquisition parameters using OPUS software suite (Bruker).

There are several operation modes available at the beamline, depending on the specific requirements of the experiment. The primary modes of operation are transmission, reflectance, transflectance, grazing incidence reflectance and attenuated total reflection (ATR). Transmission microspectroscopy generally requires the samples to be thinner than ~10 μm, whereas reflectance and transflectance measurements ideally require the sample to be highly polished or placed on a reflective substrate such as gold or on low-emissivity (low-E) slides. Grazing incidence reflectance can be applied to the study of thin films on metallic surfaces, while ATR microspectroscopy can be applied to a diverse array of samples ranging from soft gels [17], fingermarks [18] and food products [19] to single fibres [20] and composite materials [21, 22]. Details of the experimental setups for transmission and ATR measurements, which are the two main operation modes used for the investigation of insect wings, are given below.

3.1 Transmission

Prior to spectral data acquisition, the preparation of insect wing samples involves the dissection of the wing membrane into rectangular sections of approximately 5 mm × 8 mm in area using a surgical blade. The wing sections were then gently rinsed with MilliQ H_2O (resistivity of 18.2 MΩ cm^{-1}, Merck Millipore, Washington, DC, USA) and then blow-dried using 99.99% purity nitrogen gas [23–26].

For transmission measurements, the wing sections were held on a gap between an aluminium support frame using polyimide (Kapton®) tape to fix both sides of the section, before being transferred onto the sample stage of the microscope (**Figure 2A**). The IR microscope was operated with a matching 36× IR reflecting objective and condenser (NA = 0.5). In transmission mode, the synchrotron IR beam passes through a condenser located below the microscope stage and is focused into a small spot onto the wing sample (**Figure 2B**). By coupling the IR microscope to a synchrotron source, significant advantages could be gained regarding the highly collimated and highly intense synchrotron IR beam, particularly for single point FTIR microanalysis of sampling areas in the order of 3–5 μm^2. At the IRM beamline, the area of measurement per pixel on the sample is defined by the size of a single aperture located in a focal plane between the sample and detector, which can be adjusted to be as small as 4–5 μm, whereas the minimum step interval between pixels allowed for this microscope is 1 μm. Therefore, synchrotron FTIR microspectroscopy has been proven to be a powerful analytical tool for acquiring the spatially resolved distribution of chemical functional groups present across selected areas of the insect wings being analysed [14, 23–26].

3.2 Attenuated total reflection (ATR)

ATR-FTIR spectroscopy utilises the property of total internal reflection, which is established when light travelling in an optically denser medium of an ATR crystal element impinges onto the surface of a lower-index medium with an incident angle

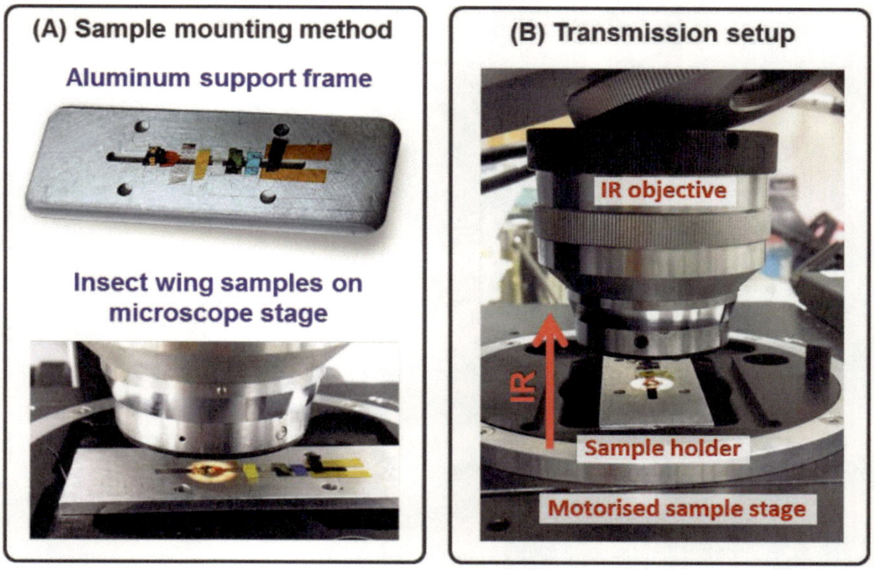

Figure 2.
(A) An aluminium support frame used as a sample holder for mounting insect wing sections prior to them being transferred onto the microscope stage and (B) schematic diagram showing the IR beam path in transmission mode for the investigation of insect wings.

greater than the critical angle [27]. Under these specific conditions, the electromagnetic field (i.e. an evanescent wave) penetrates and decays exponentially within a shallow depth of the adjacent low-index sample. Any IR-absorbing material located in contact with this interface interacts with the evanescent field resulting in an IR absorption, which is consequently transformed into an ATR-FTIR spectrum. The typical path lengths that the evanescent wave penetrates into most organic samples range from 0.2 μm to 5 μm, depending on the refractive indices of the ATR crystal and the sample, the angle of incidence and the wavelength of light [27]. As a result, the ATR-FTIR technique has been widely used for probing surface-specific molecular information of materials. The technique has been used as an alternative sampling method (i) when the sample cannot be prepared as thin sections for transmission measurements and (ii) to avoid scattering artefacts commonly presented in transflectance measurements.

Two models of ATR devices were developed for mapping measurements at the IRM beamline based on the use of an ATR hemispherical element known as the macroscopic (macro) ATR-FTIR technique [17, 28]. The technique increases the field of view and only requires a single contact point for the entire mapping measurement, preventing the cross contamination between measurement points and minimising the sample damage, which often occur in the traditional microscopic (micro) ATR-FTIR approach where repeated contacts are performed similar to a tapping mode.

While one of the ATR models was developed based on a piezo-controlled driving system [17], the so-called "hybrid" macro ATR-FTIR, which is the main ATR device used for the analysis of insect wings, was developed by modifying the cantilever arm of a standard Bruker macro ATR-FTIR unit to accept germanium (Ge) ATR elements with different facet sizes (i.e. 1 mm, 250 μm and 100 μm in diameter) to suit for different types of sample surfaces (**Figure 3A**). Coupling the synchrotron IR beam to the Ge ATR hemispherical crystal ($n_{\mathrm{Ge}} = 4$) using a high NA microscope objective essentially reduces (i) the beam focus size and (ii) the mapping step size (relative to the stage step motion) by a factor of 4, hence further improving the spatial resolution when compared to transmission or transflectance microspectroscopy. As a result, synchrotron macro ATR-FTIR measurements can be performed

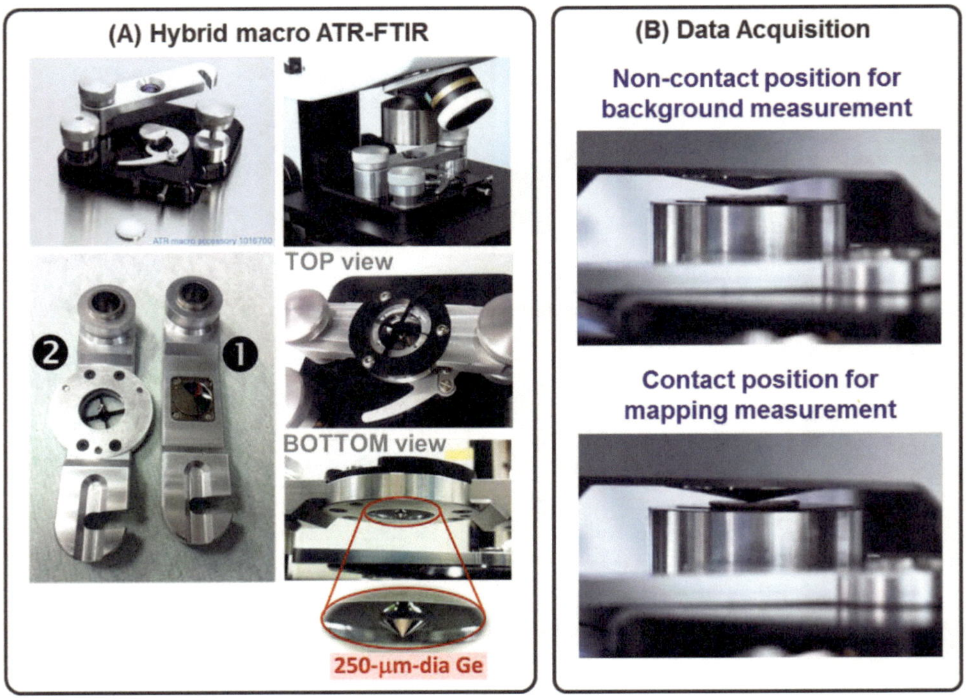

Figure 3.
(A) Hybrid macro ATR-FTIR device, showing the standard macro ATR-FTIR unit (top) and the adaptation of the cantilever arm to couple with the smaller diameter Ge micro-ATR crystal (2) compared to the original 1 mm diameter Ge macro ATR crystal (1) (bottom) and (B) data acquisition process involving background measurement of air in the noncontact position followed by mapping measurement of the wing sample in the contact position.

with aperture sizes that are 4 times larger, thereby eliminating the diffraction effects introduced by small aperture dimensions, making it ideal for high-resolution chemical analysis.

In practice, the wing section was mounted on an aluminium disc using polyimide (Kapton®) tape to hold both sides of the wing. The aluminium disc was then placed into the sample stage of the macro ATR-FTIR unit. As illustrated in **Figure 3B**, the Ge ATR crystal was subsequently brought to the focus of the synchrotron IR beam, and a background spectrum was recorded in air using 4 or 8 cm^{-1} spectral resolution and 256 co-added scans. After that, the wing sample was brought into contact with the Ge ATR crystal, and a synchrotron macro ATR-FTIR chemical map was acquired using shorter co-added scans (such as 8, 16 or 32 scans) depending on the observed absorption intensity and the quality of the contact. With the use of a Ge ATR crystal ($n_{Ge} = 4$), the area of measurement per pixel on the wing sample can be as small as 1.9 µm, and the minimum step interval between pixels achievable is 0.25 µm.

The use of synchrotron-source infrared (IR) spectroscopy is an effective method of surface characterisation to determine the chemical composition across insect wings. The intense and highly collimated light source allows sensitive detection of chemical components with high-resolution spatial mapping.

4. Case study: investigation and characterisation of insect wings

The wings of some insects exhibit unique anti-wetting, anti-biofouling, self-cleaning and bactericidal properties, primarily due to nanostructured arrays of lipids on the wing epicuticle [4–8]. These properties are of interest as a biomimetic template for the design of anti-biofouling and self-cleaning surfaces, with applications in a range of industries [9]. Additionally, the mechano-bactericidal activity of these surfaces provides a novel solution to preventing bacterial colonisation on biomedical devices and other surfaces, particularly in light of emerging concerns of antibiotic resistance. This bactericidal activity was first discovered on the wings of cicadas and shortly after dragonflies and damselflies and has since been successfully extended to synthetic analogues of the wings, with similar high aspect ratio nanoscale features [5, 29–31]. These surfaces present advantages over traditional strategies that typically rely on coatings embedded with antimicrobial agents, which can have issues of reduced efficacy over time and cytotoxic side effects and promote the development of resistance [10–13].

The wing membrane of these insects can be broken into two primary components, the procuticle and the epicuticle (**Figure 4**). These insect wings have been widely studied as strong, lightweight materials, especially in the context of aerodynamics [32–34]. The wing can be broadly defined as composed of two layers, the procuticle, which is the predominate layer and the epicuticle, which consists of a

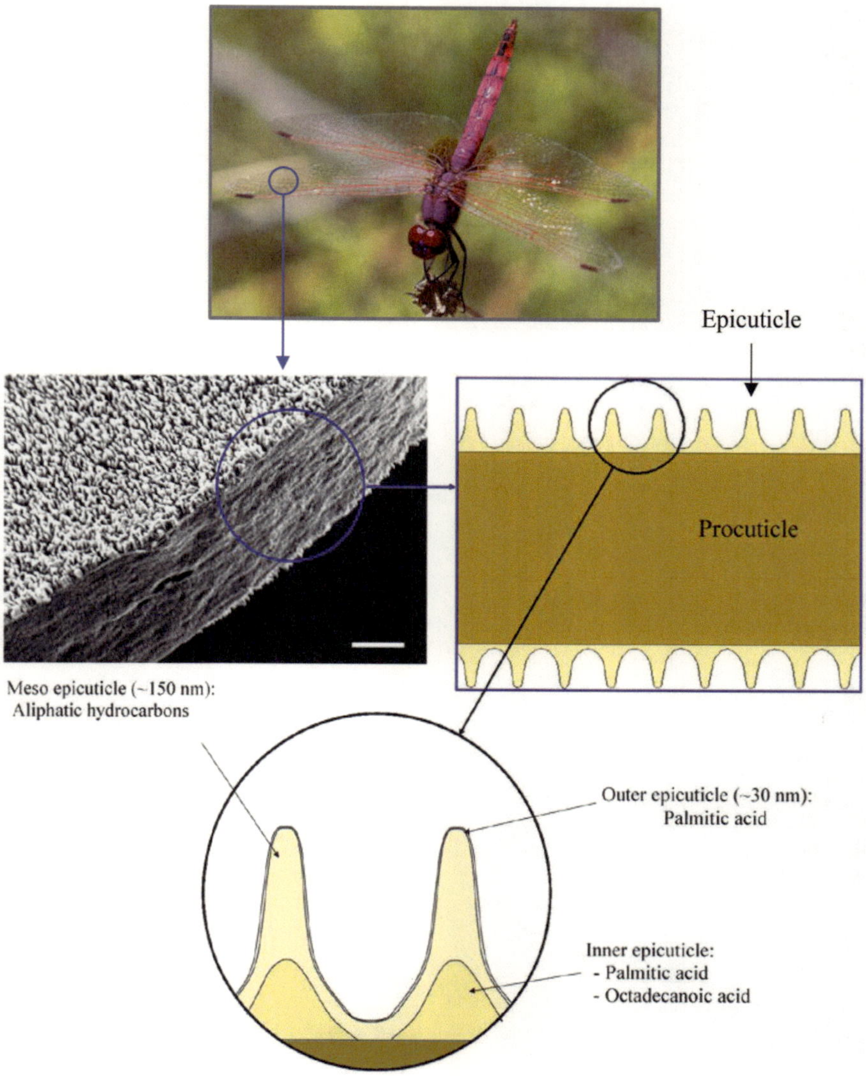

Figure 4.
Proposed model of the epicuticle of the dragonfly wing membrane. Three layers are contained within the epicuticle: the outer epicuticle, the meso epicuticle and the inner epicuticle (adapted from Ivanova et al.) [5].

thin layer of nanoscale pillars on both the dorsal and ventral sides of the procuticle and significantly contributes to the surface properties of the wings.

Synchrotron-source IR spectroscopy is useful for investigating the chemical composition of defined areas of materials. IR spectroscopy provides important information of the molecular bonding, groups and structure of chemical identities present in a sample. The extremely intense and collimated light generated enables significantly improved sensitivity, with a low signal-to-noise ratio. Furthermore, the IRM beamline at the Australian Synchrotron is coupled to a Fourier-transform infrared (FTIR) spectrometer and microscope, providing a highly focused IR beam, which enables spatially resolved chemical maps close to the diffraction limit, to be generated pixel by pixel. These properties are advantageous when characterising the chemical composition and distribution of the insect wing membrane, which is only a few microns thick, and possess lateral features on the nano-/microscale. A comparison with other surface chemistry characterisation techniques is made in **Table 1**.

Technique	Insect wing sample	Advantages	Disadvantages	Refs
Synchrotron-source IR spectroscopy	Dragonfly, cicada and damselfly wings	Performed in atmospheric conditions and suitable for sensitive samples Able to map chemical functionality across the wings with high spatial resolution	Does not provide the identity of individual chemical species (only chemical bonding)	[14, 24, 36]
X-ray photoelectron spectroscopy	Dragonfly wings, *Hemianax papuensis* 15 species of cicadas	Overall chemistry of insect wing surfaces at the top layer Capable of surface excavation to reveal the chemical depth profile	Does not provide details of chemical bonding Unable to map	[35] [37]
Raman microspectroscopy	*Drosophila* wings	Overall chemical functionality of insect wing surfaces Able to map the wing surfaces with pixel-to-pixel distance less than 1 μm	Low signal-to-noise ratio Interference by the fluorescent background of the insect wings	[38]
MALDI-TOF mass microspectroscopy	Wings of the grey flesh fly (*Neobellieria bullata*) and common fruit fly (*Drosophila melanogaster*)	Provides the direct chemical functionality of the insect wing surface Able to map chemical functionality across the insect wings	The samples are required to be completely flat Low spatial resolution Destructive to the sample	[39]

Table 1.
A comparison of different techniques to characterise the surface chemistry of different insect wings.

4.1 Transmission mode

Insect wings were cut into small sections using a sterile surgical blade. Spectra were collected in transmission mode within the 4000–800 cm^{-1} spectral region. High-resolution spectral maps of the wing samples were obtained through FTIR micro-spectroscopy at the infrared microspectroscopy (IRM) beamline at the Australian Synchrotron. These spectral maps detail spatial variations in peaks representative of lipid components (ester carbonyl C = O peak or C—H stretching peaks). A comparison of the nanostructures present on the wing membrane surface and lipid spectral maps of different cicada, dragonfly and damselfly species is presented in **Table 2**.

The wing membrane of cicadas, on both the ventral and dorsal sides, consists of a periodic hexagonal array of spherically capped nanopillars [23, 44]. Comparatively, the wing membranes of dragonflies and damselflies are covered with an irregular array of single and clustering nanopillars, which are typically taller with a larger distribution in height than those present on cicada wings.

Where provided, the 2D spectral maps of the dragonfly wings showed spatial variations in the intensity of the ester carbonyl peak and ν_{as} CH_2 stretching peak, both a characteristic of lipids. Most studies focus on the membrane of the wing, as this is primarily responsible for the unique surface properties of interest. However, Cheeseman et al. [36] use wider fields of view which encompass regions of the vein and membrane where it can be observed that the distribution of lipids was higher in the membrane sections, represented in red, when compared to the wing veins, indicated in blue [36]. Among the dragonfly and damselfly wing samples, the lipid spectral maps appear heterogenous across the surface, often resulting in microdomains of higher concentrations of lipids. This is likely to reflect the unique array of nanopillars, as indicated in the scanning electron micrographs, which tend to cluster. In comparison, the cicada sample appears more homogenous, possibly reflecting the uniform pattern of free-standing nanopillars on the epicuticle.

The average spectra from each sample contain three major band groups at 3500–3100 cm^{-1}, 3000–2800 cm^{-1} and 1800–1450 cm^{-1} which correspond to the hydroxyl, alkyl hydrocarbons and ester carbonyl groups, respectively [3, 45, 46]. The presence of C—H stretching bands, particularly ν_{as} CH_2 and ν_s CH_2, indicates long-chain aliphatic hydrocarbons typical of waxes [1, 45]. The spectra of all wings were dominated by amide I and amide II absorption bands, due to C = O bond stretching coupled to N—H bending (1695–1610 cm^{-1}) and C—N stretching coupled to N—H bending (1575–1480 cm^{-1}), respectively [3]. The presence of these amide groups is attributed to the chitin and protein components of the wings, as they represent the major structural components of the insect cuticle. The C—H stretching region (2840–3000 cm^{-1}) represents the symmetric (ν_s) and antisymmetric (ν_{as}) stretching vibrations of the CH_2 and CH_3 functional groups. The high proportion of CH_2 stretching bands in comparison to CH_3, indicates the presence of long-chain aliphatic hydrocarbons [1, 45]. To further understand the top surface layer of the wing membrane, Tobin et al. [14] subtracted spectra extracted from a point of low intensity on the ν_{as} CH_2 integration map from a point of high intensity. This was performed in collaboration with the Synchrotron Radiation Center (SRC), University of Wisconsin-Madison, performed at the IRENI Beamline. Following subtraction, the only significant peaks which remained corresponded to ν_s CH_2 and ν_{as} CH_2 vibrations in the C-H stretching region, consistent with a long-chain hydrocarbon [14]. An additional approach to identify the top surface layer has been taken in multiple studies, where by spectral measurements were performed on wings following the selective removal of the lipid components using chloroform, demonstrating a reduction of the C—H stretching bands [14, 35, 41]. The intensity of these C—H stretching bands varied between the insects (**Figure 5**). The cicada

Insect group	Species	Physical structure	Lipid spectral map[*]
Cicada			
 [14]	*Psaltoda claripennis* [23]	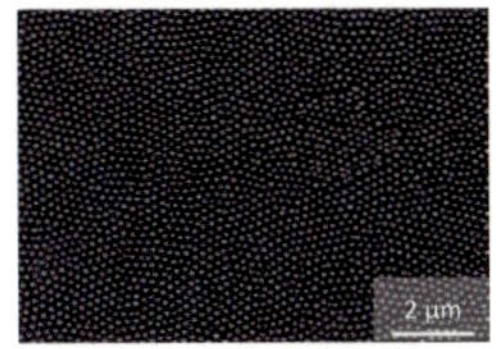	No data available
	Psaltoda claripennis[a] [14]	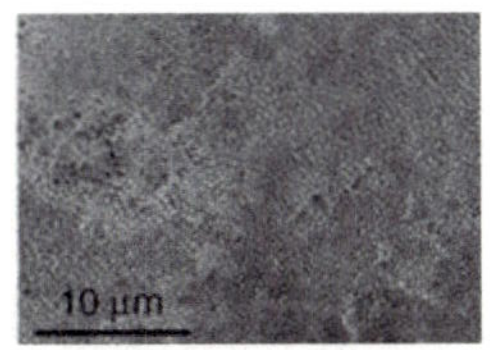	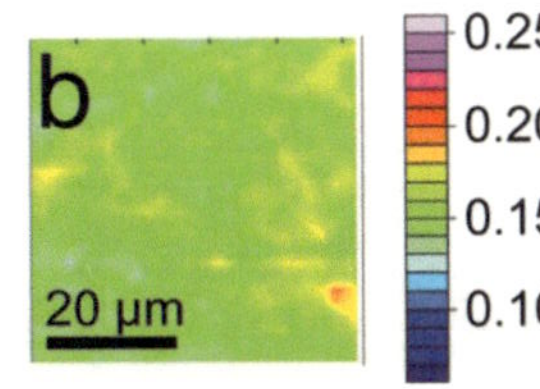
Dragonfly			
 [40]	*Austrothemis nigrescens*[b] [36]	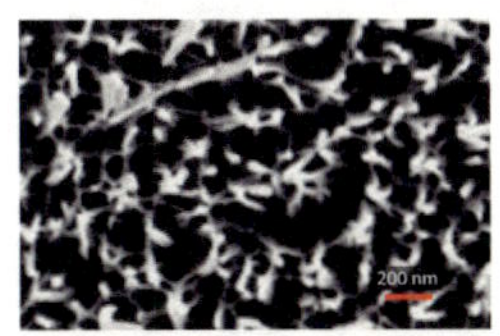	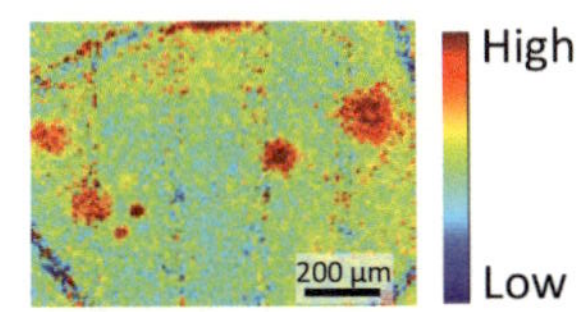
	Orthetrum chrysostigma[b] [36]	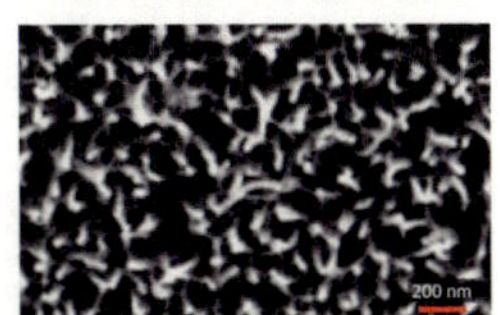	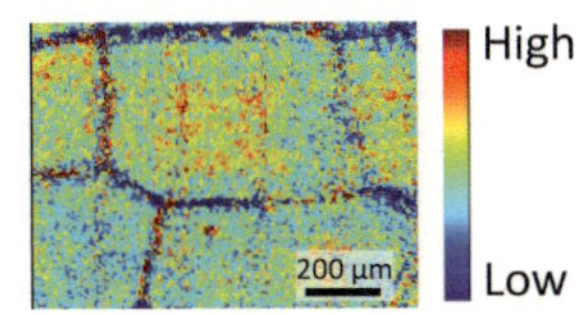

Insect group	Species	Physical structure	Lipid spectral map*
	Trithemis annulata[b] [36]	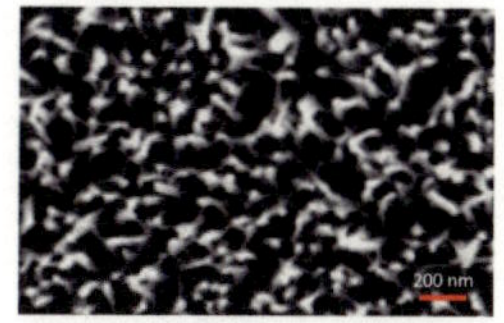	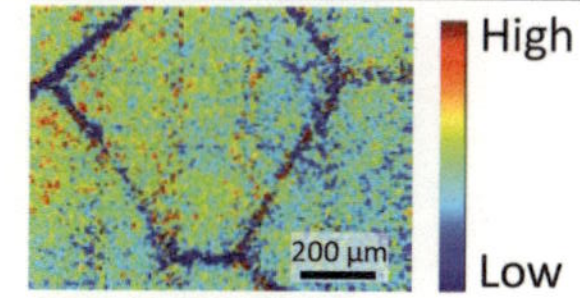
	Sympetrum fonscolombii[b] [36]	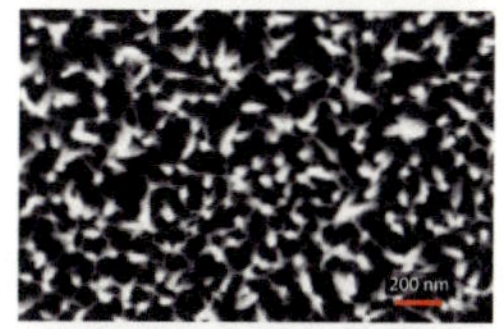	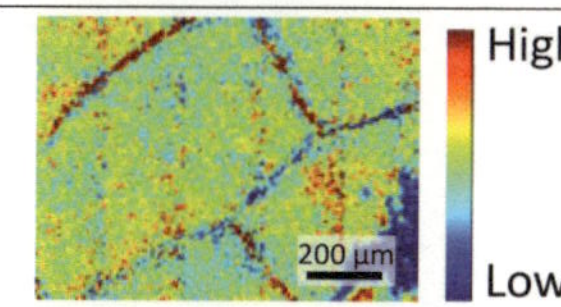
	Anax parthenope[b] [36]	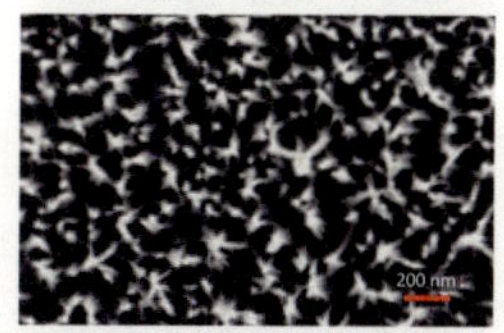	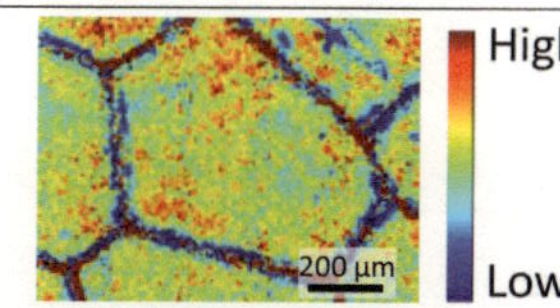
	Anax imperator[b] [36]	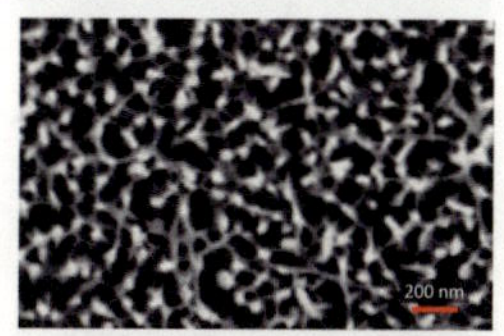	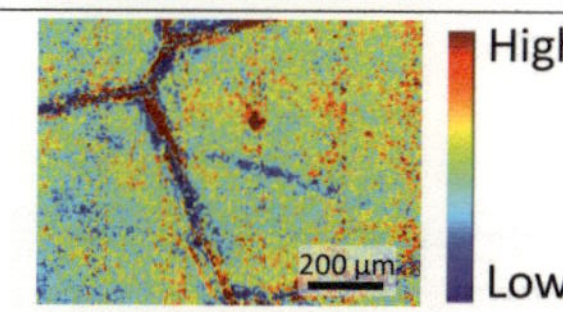
	Onychogomphus forcipatus[b] [36]	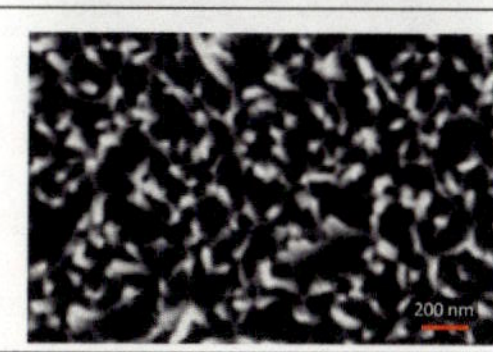	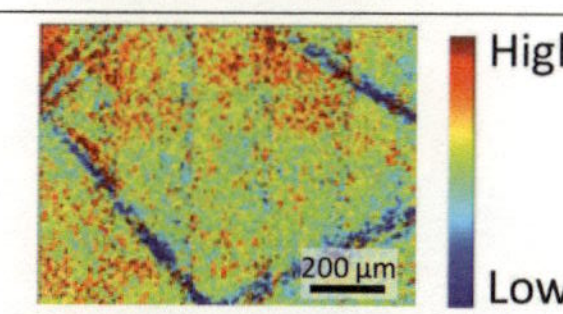

Insect group	Species	Physical structure	Lipid spectral map[*]
	Diplacodes melanopsis [7]		No data available
	Diplacodes bipunctata [7]		No data available
	Hemianax papuensis[c] [35]	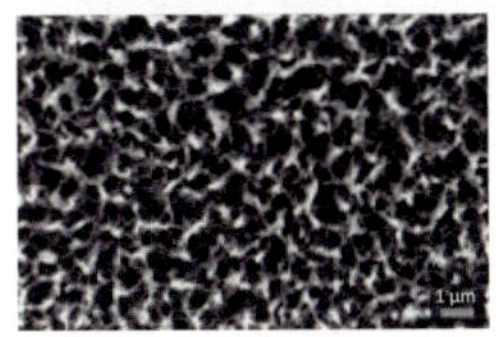	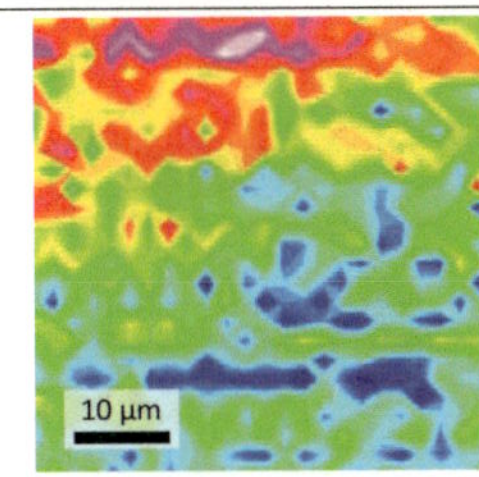
	Hemicordulia tau[c] [41]	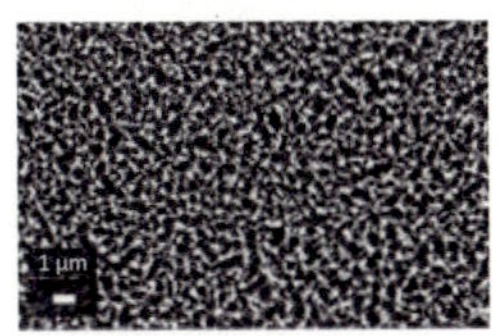	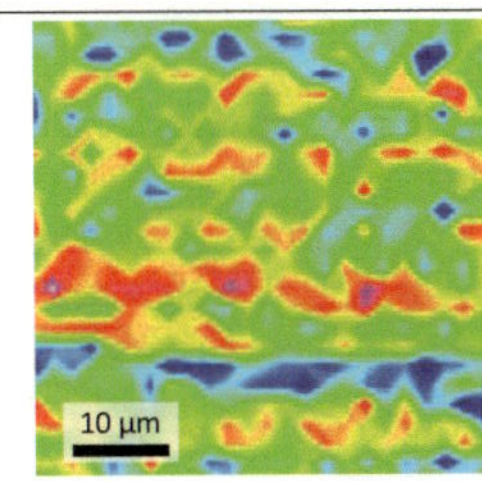

Insect group	Species	Physical structure	Lipid spectral map*
Damselfly			
 [42]	*Calopteryx haemorrhoidalis* [24]	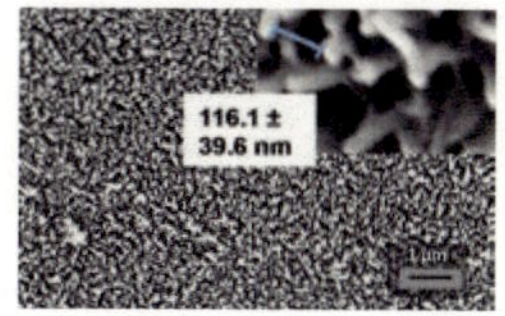	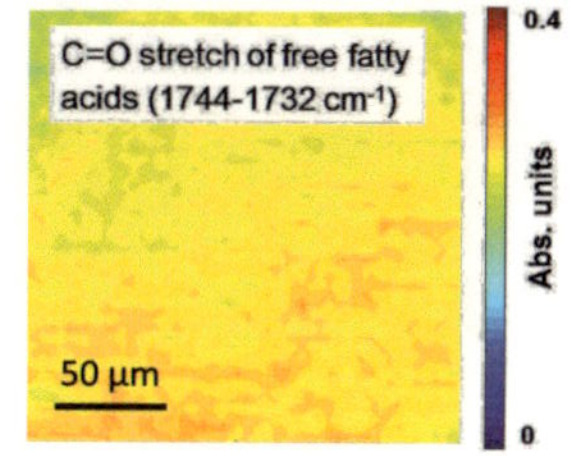
	Calopteryx haemorrhoidalis [26]		No data available
	Calopteryx haemorrhoidalis [25]	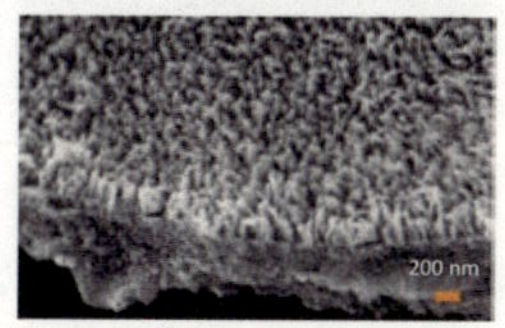	No data available
	Xanthagrion erythroneurum [7]	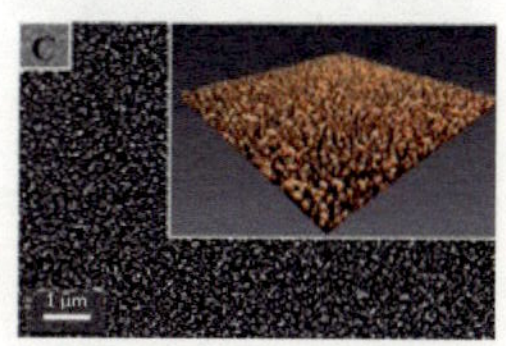	No data available

DOI: http://dx.doi.org/10.5772/intechopen.84591

Insect group	Species	Physical structure	Lipid spectral map*
	Ischnura heterosticta [7]		No data available
	Ischnura heterosticta [43]	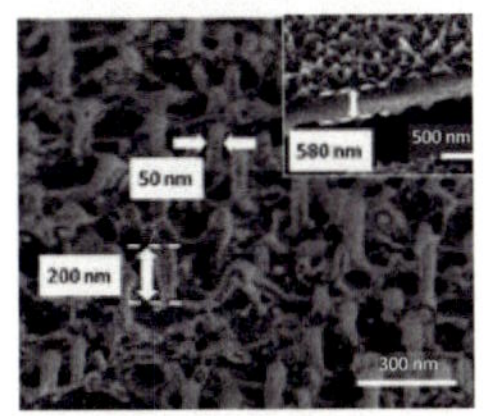	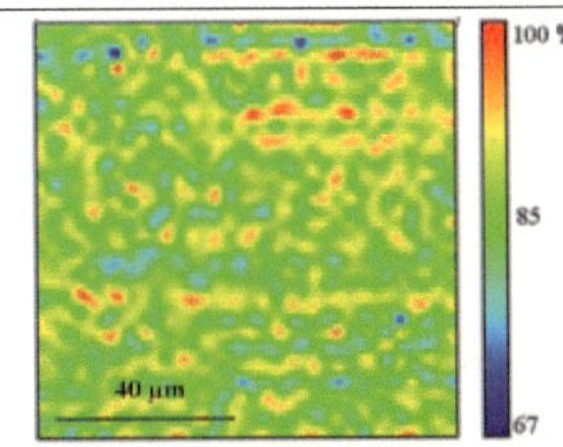

With the exception of the second entry (Psaltoda claripennis), the lipid spectral maps were generated by integrating the ester carbonyl peak, approximately in the wavelength range 1750–1720 cm^{-1}, which is representative of the C = O stretch of esters

[a]*Lipid spectral map generated by the absorbance under the ν_{as} CH_2: 2931–2913 cm^{-1}*

[b]*Spectral data in Cheeseman et al. [36] was collected using an offline Globar™ IR source at the IRM beamline at the Australian Synchrotron*

[c]*Lipid spectral maps for Hemianax papuensis and Hemicordulia tau are unpublished and were generated by integrating the ester carbonyl peak, approximately in the wavelength range 1750–1720 cm^{-1}*

Table 2.
A comparison of the nanostructures present on the wing membrane surface of different cicada, dragonfly and damselfly species and high-resolution lipid spectral maps generated using IR microspectroscopy.

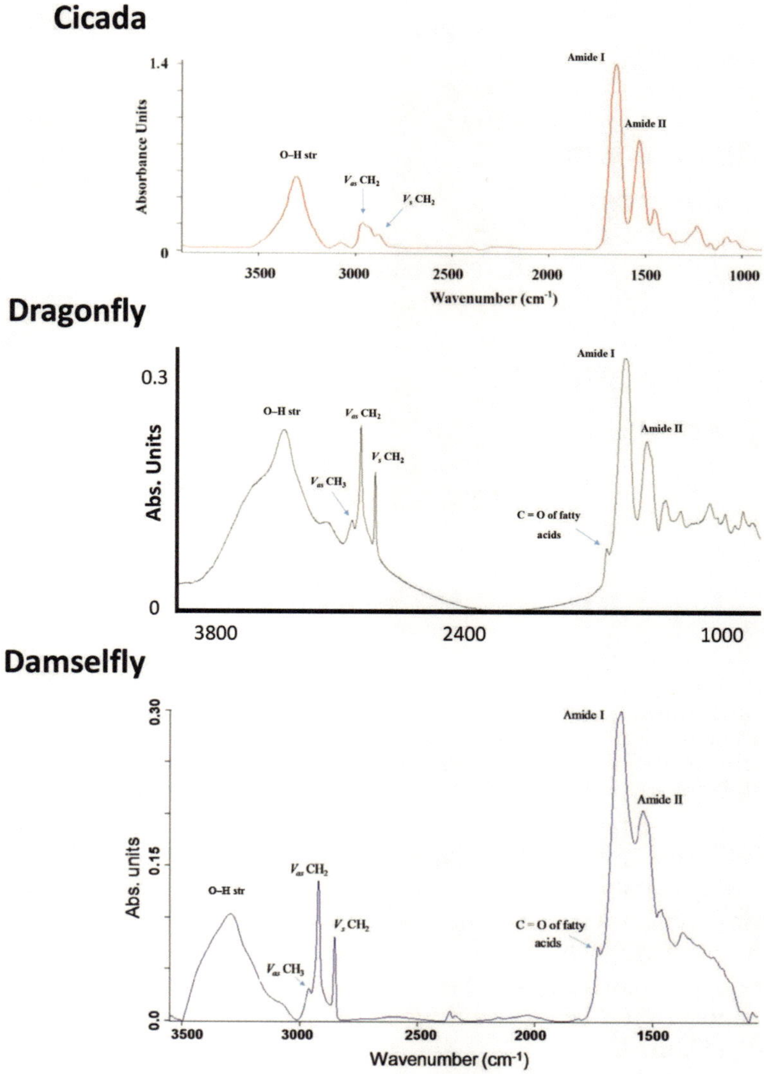

Figure 5.
Average IR spectra of an area of the wing membrane of a species of cicada (Psaltoda claripennis), (adapted from Ivanova et al. [23]), reprinted with permission from John Wiley and Sons, dragonfly (Austrothemis nigrescens) (adapted from Cheeseman et al. [36]) and damselfly (Calopteryx haemorrhoidalis) (adapted from Truong et al. [24]), both reprinted with permission from Springer.

wing sample demonstrated comparatively lower peaks in the C—H stretching region than the dragonfly and damselfly wing samples, which may reflect the reduced height of the waxy nanopillars on the cicada (~200 nm) compared to the dragonfly (~200–300 nm) and damselfly (~400–500 nm).

4.2 Transmission versus attenuated total reflectance

In addition to transmission mode, there is also attenuated total reflectance (ATR) mode available at the Australian Synchrotron. This mode utilises a Ge crystal with a high refractive index, through an in-house-developed device, generating an evanescent wave penetrating only the top surface layer of the sample, as described above. The high refractive index promotes detection of molecular bonding in very small concentrations in the sample. This mode is advantageous to transmission for thick or opaic samples, where transmission is not achievable or in circumstances

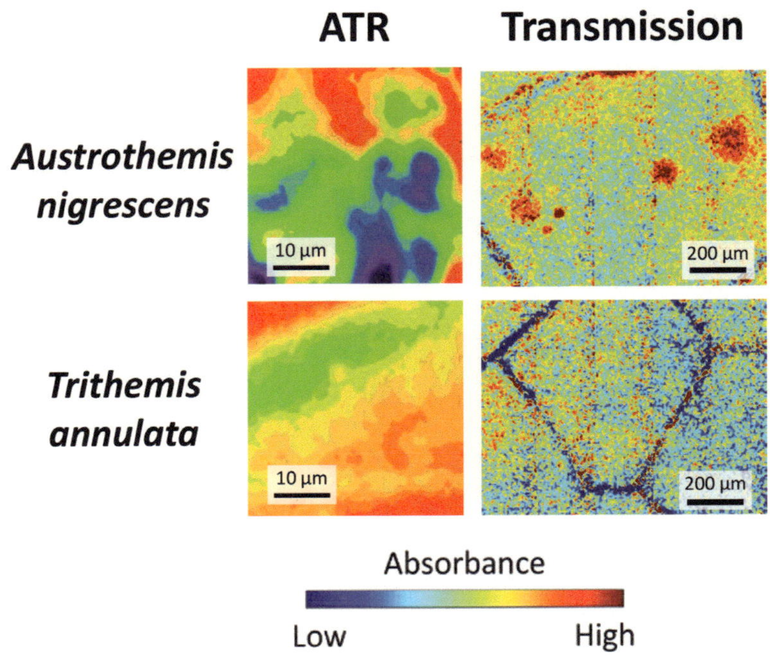

Figure 6.
Comparison of the distribution of lipid components of the wing membrane of two dragonflies, A. nigrescens and T. annulata, acquired using ATR (unpublished) and transmission (adapted from Cheeseman et al. [36]) mode IR spectroscopy at the Australian Synchrotron. Lipid spectral maps were generated by integrating the ester carbonyl peak, in the wavelength range 1750–1720 cm^{-1}, which is representative of the C = O stretch of esters.

when only the top surface of the sample is of interest. In the case of characterising insect wing membranes, ATR mode provides a complimentary technique to transmission, generating high-resolution spectral maps of only the top surfaces (primarily epicuticle), which is of interest as the primary contributor to the unique anti-wetting, anti-biofouling and bactericidal properties. A comparison of spectral maps generated from ATR and transmission mode of the wings of two dragonflies is presented in **Figure 6**.

The high spatial resolution of ATR IR spectroscopy demonstrates the lateral distributions of lipids/waxes on the micron scale (**Figure 6**). Additionally, because the light source only partially penetrates the surface in ATR mode, the spectral map more closely represents the lipid distribution on the wing epicuticle, composed of an irregular array of lipid nanopillars. However, there is also a compromise as structural details of the procuticle are not included and the total imaging area is limited to the facet size of the crystal. Additionally, because ATR mode requires physical contact between the crystal and the sample surface, image quality is affected by the topography of the surface, with issues arising with non-flat surfaces. For example, it is not possible to image both the vein and membrane sections of the dragonfly wings due to the large height difference between the membrane and vein surface.

5. Conclusions

The use of synchrotron-source infrared (IR) spectroscopy is an effective method of surface characterisation to determine the chemical composition across insect wings. The intense and highly collimated light source allows sensitive detection of chemical components with high-resolution spatial mapping. Such sensitive surface characterisation is useful for insect wings, as well as other natural materials with unique surface properties, which may have important applications as templates for biomedical, environmental or industrially relevant materials.

Acknowledgements

This research was undertaken in part on the IRM beamline at Australian Synchrotron (Victoria, Australia), part of ANSTO. The development of the macro ATR-FTIR device mentioned here was funded as part of the Science Projects at the Australian Synchrotron and has been made available to users of the IRM beamline. The authors wish to acknowledge Mr. Alan Easdon from the Australian Synchrotron, for his substantial contribution to the design and mechanical works associated with the development of this device.

Conflict of interest

The authors declare no conflicts of interest in this work.

Author details

Samuel Cheeseman[1], Vi Khanh Truong[1], Jitraporn Vongsvivut[2], Mark J. Tobin[2], Russell Crawford[1] and Elena P. Ivanova[1*]

1 School of Science, College of Science, Engineering and Health, RMIT University, Melbourne, Australia

2 Infrared Microspectroscopy (IRM) Beamline, Australian Synchrotron, ANSTO, Clayton, Victoria, Australia

*Address all correspondence to: elena.ivanova@rmit.edu.au

References

[1] Coates J. Interpretation of infrared spectra, a practical approach. In: Meyers RA, McKelvy ML, editors. Encyclopedia of Analytical Chemistry. Chichester, England: Wiley & Sons; 2000. pp. 10815-10837

[2] Vongsvivut J, Heraud P, Gupta A, Puri M, McNaughton D, Barrow CJ. FTIR microspectroscopy for rapid screening and monitoring of polyunsaturated fatty acid production in commercially valuable marine yeasts and protists. The Analyst. 2013;**138**(20):6016-6031. DOI: 10.1039/c3an00485f

[3] Barth A. Infrared spectroscopy of proteins. Biochimica et Biophysica Acta (BBA)-Bioenergetics. 2007;**1767**(9):1073-1101. DOI: 10.1016/j.bbabio.2007.06.004

[4] Bandara CD, Singh S, Afara IO, Wolff A, Tesfamichael T, Ostrikov K, et al. Bactericidal effects of natural nanotopography of dragonfly wing on *Escherichia coli*. ACS Applied Materials & Interfaces. 2017;**9**(8):6746-6760. DOI: 10.1021/acsami.6b13666

[5] Ivanova EP, Hasan J, Webb HK, Gervinskas G, Juodkazis S, Truong VK, et al. Bactericidal activity of black silicon. Nature Communications. 2013:42838-42844. DOI: 10.1038/ncomms3838

[6] Mainwaring DE, Nguyen SH, Webb H, Jakubov T, Tobin M, Lamb RN, et al. The nature of inherent bactericidal activity: Insights from the nanotopology of three species of dragonfly. Nanoscale. 2016;**8**(12):6527-6534. DOI: 10.1039/C5NR08542J

[7] Nguyen SH, Webb HK, Hasan J, Tobin MJ, Mainwaring DE, Mahon PJ, et al. Wing wettability of odonata species as a function of quantity of epicuticular waxes. Vibrational Spectroscopy. 2014;**75**(Supplement C):173-177. DOI: 10.1016/j.vibspec.2014.07.006

[8] Song Ha N, Hayden KW, Peter JM, Russell JC, Elena PI. Natural insect and plant micro−/nanostructured surfaces: An excellent selection of valuable templates with superhydrophobic and self-cleaning properties. Molecules. 2014;**19**(9):13614-13630. DOI: 10.3390/molecules190913614

[9] Gangadoo S, Chandra S, Power A, Hellio C, Watson GS, Watson JA, et al. Biomimetics for early stage biofouling prevention: Templates from insect cuticles. Journal of Materials Chemistry B. 2016;**4**(34):5747-5754. DOI: 10.1039/C6TB01642A

[10] Tiller JC, Liao C-J, Lewis K, Klibanov AM. Designing surfaces that kill bacteria on contact. Proceedings of the National Academy of Sciences of the United States of America. 2001;**98**(11):5981-5985. DOI: 10.1073/pnas.111143098

[11] Pacor S, Giangaspero A, Bacac M, Sava G, Tossi A. Analysis of the cytotoxicity of synthetic antimicrobial peptides on mouse leucocytes: Implications for systemic use. Journal of Antimicrobial Chemotherapy. 2002;**50**(3):339-348. DOI: 10.1093/jac/dkf141

[12] Mijnendonckx K, Leys N, Mahillon J, Silver S, Van Houdt R. Antimicrobial silver: Uses, toxicity and potential for resistance. Biometals. 2013;**26**(4):609-621. DOI: 10.1007/s10534-013-9645-z

[13] Stapleton PD, Taylor PW. Methicillin resistance in *Staphylococcus aureus*: Mechanisms and modulation. Science Progress. 2002;**85**(1):57-72. DOI: 10.3184/003685002783238870

[14] Tobin MJ, Puskar L, Hasan J, Webb HK, Hirschmugl CJ, Nasse MJ,

et al. High-spatial-resolution mapping of superhydrophobic cicada wing surface chemistry using infrared microspectroscopy and infrared imaging at two synchrotron beamlines. Journal of Synchrotron Radiation. 2013;**20**(3):482-489. DOI: 10.1107/S0909049513004056

[15] Caine S, Heraud P, Tobin MJ, McNaughton D, Bernard CCA. The application of Fourier transform infrared microspectroscopy for the study of diseased central nervous system tissue. NeuroImage. 2012;**59**(4):3624-3640. DOI: 10.1016/j.neuroimage.2011.11.033

[16] Miller LM, Dumas P. Chemical imaging of biological tissue with synchrotron infrared light. Biochimica et Biophysica Acta (BBA)—Biomembranes. 2006;**1758**(7):846-857. DOI: 10.1016/j.bbamem.2006.04.010

[17] Benbow NL, Webber JL, Pawliszak P, Sebben DA, Ho TTMV, Vongsvivut J, et al. A novel soft contact piezo-controlled liquid cell for probing polymer films under confinement using synchrotron FTIR microspectroscopy. Scientific Reports. 2019. In press

[18] Dorakumbura BN, Boseley RE, Becker T, Martin DE, Richter A, Tobin M, et al. Revealing the spatial distribution of chemical species within latent fingermarks using vibrational spectroscopy. The Analyst. 2018;**143**:4027-4039. DOI: 10.1039/C7AN01615H

[19] Timilsena YP, Vongsvivut J, Tobin MJ, Adhikari R, Barrow C, Adhikari B. Investigation of oil distribution in spray-dried chia seed oil microcapsules using synchrotron-FTIR microspectroscopy. Food Chemistry. 2019:275457-275466. DOI: 10.1016/j.foodchem.2018.09.043

[20] Ryu M, Balčytis A, Wang X, Vongsvivut J, Hikima Y, Li J, et al. Orientational mapping augmented sub-wavelength hyper-spectral imaging of silk. Scientific Reports. 2017;**7**(1):7419. DOI: 10.1038/s41598-017-07502-3

[21] Nunna S, Creighton C, Fox BL, Naebe M, Maghe M, Tobin MJ, et al. The effect of thermally induced chemical transformations on the structure and properties of carbon fibre precursors. Journal of Materials Chemistry A. 2017;**5**(16):7372-7382. DOI: 10.1039/C7TA01022B

[22] Vongsvivut J, Truong VK, Al Kobaisi M, Maclaughlin S, Tobin MJ, Crawford RJ, et al. Synchrotron macro ATR-FTIR microspectroscopic analysis of silica nanoparticle-embedded polyester coated steel surfaces subjected to prolonged UV and humidity exposure. PLoS One. 2017;**12**(12):e0188345. DOI: 10.1371/journal.pone.0188345

[23] Ivanova EP, Hasan J, Webb HK, Truong VK, Watson GS, Watson JA, et al. Natural bactericidal surfaces: Mechanical rupture of *Pseudomonas aeruginosa* cells by cicada wings. Small. 2012;**8**(16):2489-2494. DOI: 10.1002/smll.201200528

[24] Truong VK, Geeganagamage NM, Baulin VA, Vongsvivut J, Tobin MJ, Luque P, et al. The susceptibility of *Staphylococcus aureus* CIP 65.8 and *Pseudomonas aeruginosa* ATCC 9721 cells to the bactericidal action of nanostructured *Calopteryx haemorrhoidalis* damselfly wing surfaces. Applied Microbiology and Biotechnology. 2017;**101**(11):4683-4690. DOI: 10.1007/s00253-017-8205-9

[25] Stuhr S, Truong VK, Vongsvivut J, Senkbeil T, Yang Y, Al Kobaisi M, et al. Structure and chemical organization in damselfly *Calopteryx haemorrhoidalis* wings: A spatially resolved FTIR and XRF analysis with synchrotron radiation. Scientific Reports. 2018;**8**(1):8413. DOI: 10.1038/s41598-018-26563-6

[26] Truong VK, Vongsvivut J, Mahanamanam N, Tobin MJ, Luque P, Baulin V, et al. Study of melanin localization in the mature male *Calopteryx haemorrhoidalis* damselfly wings. Journal of Synchrotron Radiation. 2018;**25**(3):874-877. DOI: 10.1107/S1600577518004460

[27] Harrick NJ. Internal Reflection Spectroscopy. NY: John Wiley & Sons; 1967

[28] Truong VK, Stefanovic M, Maclaughlin S, Tobin M, Vongsvivut J, Al Kobaisi M, et al. The evolution of silica nanoparticle-polyester coatings on surfaces exposed to sunlight. Journal of Visualized Experiments. 2016:e54309

[29] Bhadra CM, Truong VK, Pham VT, Al Kobaisi M, Seniutinas G, Wang JY, et al. Antibacterial titanium nano-patterned arrays inspired by dragonfly wings. Scientific Reports. 2015:516817. DOI: 10.1038/srep16817

[30] Hasan J, Raj S, Yadav L, Chatterjee K. Engineering a nanostructured "super surface" with superhydrophobic and superkilling properties. RSC Advances. 2015;**5**(56):44953-44959. DOI: 10.1039/C5RA05206H

[31] Linklater DP, Nguyen HKD, Bhadra CM, Juodkazis S, Ivanova EP. Influence of nanoscale topology on bactericidal efficiency of black silicon surfaces. Nanotechnology. 2017;**28**(24):245301. DOI: 10.1088/1361-6528/aa700e

[32] Appel E, Heepe L, Lin C-P, Gorb SN. Ultrastructure of dragonfly wing veins: Composite structure of fibrous material supplemented by resilin. Journal of Anatomy. 2015;**227**(4): 561-582. DOI: 10.1111/joa.12362

[33] Sivasankaran PN, Ward TA, Viyapuri R, Johan MR. Static strength analysis of dragonfly inspired wings for biomimetic micro aerial vehicles. Chinese Journal of Aeronautics. 2016;**29**(2):411-423. DOI: 10.1016/j.cja.2016.02.007

[34] Jongerius SR, Lentink D. Structural analysis of a dragonfly wing. Experimental Mechanics. 2010;**50**(9):1323-1334. DOI: 10.1007/s11340-010-9411-x

[35] Ivanova EP, Nguyen SH, Webb HK, Hasan J, Truong VK, Lamb RN, et al. Molecular organization of the nanoscale surface structures of the dragonfly *Hemianax papuensis* wing epicuticle. PLoS One. 2013;**8**(7):e67893. DOI: 10.1371/journal.pone.0067893

[36] Cheeseman S, Owen S, Truong VK, Meyer D, Ng SH, Vongsvivut J, et al. Pillars of life: Is there a relationship between lifestyle factors and the surface characteristics of dragonfly wings? ACS Omega. 2018;**3**(6):6039-6046. DOI: 10.1021/acsomega.8b00776

[37] Sun M, Watson GS, Zheng Y, Watson JA, Liang A. Wetting properties on nanostructured surfaces of cicada wings. Journal of Experimental Biology. 2009;**212**(19):3148-3155. DOI: 10.1242/jeb.033373

[38] Valmalette JC, Raad H, Qiu N, Ohara S, Capovilla M, Robichon A. Nano-architecture of gustatory chemosensory bristles and trachea in *Drosophila* wings. Scientific Reports. 2015:514198. DOI: 10.1038/srep14198

[39] Vrkoslav V, Muck A, Cvačka J, Svatoš A. MALDI imaging of neutral cuticular lipids in insects and plants. Journal of the American Society for Mass Spectrometry. 2010;**21**(2):220-231. DOI: 10.1016/j.jasms.2009.10.003

[40] Hort J, Hort F. *Austrothemis nigrescens* [Image]. 2011. Available from: http://eol.org/data_objects/31719003 [Accessed: 01/11/2018]

[41] Nguyen SHT, Webb HK, Hasan J, Tobin MJ, Crawford RJ, Ivanova EP.

Dual role of outer epicuticular lipids in determining the wettability of dragonfly wings. Colloids and Surfaces B: Biointerfaces. 2013:106126-106134. DOI: 10.1016/j.colsurfb.2013.01.042

[42] Hampshire G. *Calopteryx haemorrhoidalis* [Image]. 2013. Available from: http://eol.org/pages/4139774/overview [Accessed: 01/11/2018]

[43] Hasan J, Webb HK, Truong VK, Watson GS, Watson JA, Tobin MJ, et al. Spatial variations and temporal metastability of the self-cleaning and superhydrophobic properties of damselfly wings. Langmuir. 2012;**28**(50):17404-17409. DOI: 10.1021/la303560w

[44] Watson GS, Myhra S, Cribb BW, Watson JA. Putative functions and functional efficiency of ordered cuticular nanoarrays on insect wings. Biophysical Journal. 2008;**94**(8):3352-3360. DOI: 10.1529/biophysj.107.109348

[45] Coury C, Dillner AM. A method to quantify organic functional groups and inorganic compounds in ambient aerosols using attenuated total reflectance FTIR spectroscopy and multivariate chemometric techniques. Atmospheric Environment. 2008;**42**(23):5923-5932. DOI: 10.1016/j.atmosenv.2008.03.026

[46] Lasch P, Boese M, Pacifico A, Diem M. FT-IR spectroscopic investigations of single cells on the subcellular level. Vibrational Spectroscopy. 2002;**28**(1):147-157. DOI: 10.1016/S0924-2031(01)00153-9